HD 4461 .M85 2009

Mullin, Megan, 1973-

Governing the tap : special district governance and the new local politics of water

Governing the Tap

D1602423

WILLOW INTERNATIONAL LIBRARY

American and Comparative Environmental Policy
Sheldon Kamieniecki and Michael E. Kraft, series editors

A complete list of books published in the American and Comparative Environmental Policy series appears at the back of the book.

Governing the Tap
Special District Governance and the New Local Politics of Water

Megan Mullin

The MIT Press
Cambridge, Massachusetts
London, England

© 2009 Massachusetts Institute of Technology

All rights reserved. No part of this book may be reproduced in any form by any electronic or mechanical means (including photocopying, recording, or information storage and retrieval) without permission in writing from the publisher.

For information about special quantity discounts, please email special_sales@mitpress.mit.edu

This book was set in Sabon on 3B2 by Asco Typesetters, Hong Kong. Printed and bound in the United States of America.

Library of Congress Cataloging-in-Publication Data

Mullin, Megan, 1973–
Governing the tap : special district governance and the new local politics of water / Megan Mullin.
 p. cm. — (American and comparative environmental policy)
Includes bibliographical references and index.
ISBN 978-0-262-01313-0 (hardcover : alk. paper) — ISBN 978-0-262-51297-8 (pbk. : alk. paper)
1. Water utilities—United States. 2. Water districts—United States. I. Title.
HD4461.M85 2009
363.6′10973—dc22
                                                            2008044250

10 9 8 7 6 5 4 3 2 1

For Waugh

Contents

Series Foreword ix
Acknowledgments xi

1 Introduction 1

2 A Conditional Theory of Specialized Governance 25

3 Private Costs and Public Benefits in Local Public Services 55

4 Distributing the Price of Growth 81

5 Boundaries and the Incentive to Act Alone 103

6 Fighting over Land and Water: Venues in Local Growth Disputes 123

7 Specialization and Fragmentation in American Local Governance 175

Appendix 1 Explanation of Data and Model, Chapter 3 195

Appendix 2 Explanation of Data and Model, Chapter 4 203

Appendix 3 Explanation of Data and Model, Chapter 5 209

Notes 213
References 229
Index 253

Series Foreword

In response to the increased volume of pollutants entering surface water and groundwater, thereby threatening the quality of community drinking water, the U.S. Congress passed the Safe Drinking Water Act in 1974 and major amendments in 1986 and 1996. Since then, the U.S. Environmental Protection Agency has identified a large number and wide variety of dangerous pollutants that can contaminate drinking water, and it has successfully led the effort to protect the quality of the nation's drinking water. The ongoing monitoring and regulation of tap water is no small task given the existence of more than 160,000 public-water systems throughout the United States as well as the wide range of local conditions that can influence water quality in American communities.

In addition to concern over the quality of the nation's drinking water, increasing demand on local drinking water resources has led to water shortages during periods of drought. Although primarily a problem in western states in the past, recent population growth and migration in other regions of the country have left water systems vulnerable to serious shortages of drinking water. Such incidents have prompted many communities to enact water-conservation measures and regulations and even building moratoria in order to guarantee the supply of scarce drinking water. Protecting both the quality and the quantity of a community's water supply is a growing challenge for the local institutions governing drinking water in the United States. Many of these institutions are specialized water districts that have little public visibility but make decisions about the most essential community resource.

Megan Mullin's book provides an insightful, in-depth analysis of the consequences of specialization and fragmentation for local policymaking by using decision making about local drinking water as an example. As she argues in this well-written volume, local drinking water policy is an ideal case because some of the first special districts focused their efforts on drinking water. Water districts were later used as a model for the spread of specialized governance into other local government functions (e.g., public health and fire protection). In addition, the investigation of local water-governing bodies is important now because of the rise of the new local politics of water involving both the quality and the quantity of water supplies. As Mullin points out, a system of specialized governance narrows each government's decision authority to one issue or a narrow set of issues, thereby raising a number of critical questions. For example, how does such a system affect the kinds of policies that governments make, and does it influence their responsiveness to citizens' preferences and demands? Her analysis and intriguing findings have important implications for local water policy in particular and for specialized governance in general.

The book illustrates well the goals of the MIT Press American and Comparative Environmental Policy series. We encourage work that examines a broad range of environmental policy issues. We are particularly interested in volumes that incorporate interdisciplinary research and focus on the linkages between public policy and environmental problems as well as on issues both within the United States and in cross-national settings. We welcome contributions that analyze the policy dimensions of relationships between humans and the environment from either a theoretical or empirical perspective. At a time when environmental policies are increasingly seen as controversial and new approaches are being implemented widely, we especially encourage studies that assess policy successes and failures, evaluate new institutional arrangements and policy tools, and clarify new directions for environmental politics and policy. The books in this series are written for a wide audience that includes academics, policymakers, environmental scientists and professionals, business and labor leaders, environmental activists, and students concerned with environmental issues. We hope they contribute to public understanding of environmental problems, issues, and policies of concern today and suggest promising actions for the future.

Sheldon Kamieniecki, *University California, Santa Cruz*
Michael E. Kraft, *University of Wisconsin, Green Bay*
American and Comparative Environmental Policy series editors

Acknowledgments

This project took shape during my time at the University of California, Berkeley, and I owe many debts to that institution and the people there. Judy Gruber lit the spark for the project by introducing me to the thorny democratic dilemmas inherent in the delivery of public services. I am grateful to Ray Wolfinger, Margaret Weir, John Ellwood, and especially Bruce Cain for their guidance through the early stages of the research. Thanks also to Merrill Shanks, Todd LaPorte, Henry Brady, Michael Hanemann, the Graduate Workshop in American Politics, and the talented librarians and staff at the Institute of Governmental Studies and the Water Resources Center Archive. The Temple University Department of Political Science provided valuable resources and support to help me finish the book. I am particularly grateful to Gary Mucciaroni and Chris Wlezien for their useful insight on chapter drafts, and to Kevin Arceneaux, who read the entire manuscript.

Portions of the book have been presented at Georgetown University, Temple University, the University of Michigan, MIT, the Harris School at the University of Chicago, Florida State University, the University of Kentucky, the University of Kansas, and annual meetings of the American Political Science Association and the Midwest Political Science Association. Discussants and audience members at all of these talks provided feedback that helped shape my arguments. Richard Dilworth and Alex Holzman made excellent suggestions at a critical moment in the book's development. In addition, I owe great thanks to Ellen Hanak for sharing data and her water expertise. Richard Krop at the Cadmus Group and Kurt Keeley and Patrick McElhany from the American Water Works Association also supplied data and answered many questions.

All of the people I interviewed for this project were generous with their time and thoughtful about the politics of retail water service. Many went out of their way to help me locate archival documents. Yphtach Lelkes, Joshua Weikert, Jay Jennings, Christina Wong, and Elizabeth Mattiuzzi provided stellar research assistance. Financial support came from a National Science Foundation dissertation grant (SES-0315293) as well as from grants from the Department of Political Science, the Graduate Division, and the Institute of Governmental Studies at the University of California, Berkeley. My editor at MIT Press, Clay Morgan, kindly and efficiently guided me through the publishing process and selected three reviewers whose recommendations improved the manuscript. I also appreciate the support this project has received from the two series editors, Sheldon Kamieniecki and Michael Kraft. Much of chapter 3 appeared in 2008 in the *American Journal of Political Science*, and I thank the publisher for permission to use the material here. Ken and Soni Wright helped more than they probably know by providing a space to write where it was impossible to forget the importance of water governance. Among the many colleagues and friends who have advised and encouraged me, Dorothy Daley and Patrick Egan deserve special mention for their wisdom, patience, and very good humor.

A first book feels like the culmination of a lifetime of learning. I thank my family—Diane Mullin, the dearly missed Richard Mullin, Bea Holt, Ken Wright, Soni Wright, and my many fabulous siblings—for encouraging my inquisitiveness. Most of all, I am grateful to Waugh. It would take another book, or several, to say why.

1
Introduction

Governance of American communities is becoming more specialized. Independent special districts play a growing role in providing a wide array of local services, with the consequence that most households now fall within a multiplicity of local jurisdictions. Special districts, sometimes called *public authorities*, are autonomous governments that can perform almost any of the functions of a city or county. Over the past fifty years, their number has more than tripled, making the special district the most common form of local government in the United States. Because each special district has only a limited purpose, a system of specialized governance fragments authority over a community's public services among multiple independent institutions.

This book examines the consequences of specialization and fragmentation for local policymaking. Specialization is a common method for managing growth in the size and complexity of a political system (Dahl and Tufte 1973). Legislative districts allow public officials to specialize territorially, and governments establish bureaucracies, legislative committee systems, and independent commissions to organize their work along functional lines and to promote issue expertise. Special districts represent the next step in specialization: the formation of autonomous governments with jurisdictions defined by function as well as by geography. A system of specialized governance narrows each government's decision authority to a single issue or a narrow set of issues. Several questions can be asked about such a system: How does it influence the kinds of decisions that governments make? Does it affect their responsiveness to the preferences of constituents? Do certain interests in a community enjoy a particular advantage in one kind of institutional setting over another? And to what extent are specialized governments able to coordinate their activities in order to address complex, regional policy challenges?

The answers to these questions have important consequences for where and how we live. As special districts proliferate, they absorb more functions from traditional cities and counties, and they take on added responsibility for providing essential public goods. In assigning the location of hospitals and firehouses or in treating drinking water for toxic contaminants, special districts help protect public health and safety. They boost property values when they install sewers in a neighborhood or reinforce a levee. Their choices when allocating resources for parks, libraries, and public transit have significant impact on people's job opportunities and quality of life. Through their control over infrastructure and public services, special districts can help define our physical communities, guide their growth, and influence their composition. At the same time, the crosscutting jurisdictional boundaries that emerge in a system of specialized governance may divide political communities and erode perceptions of common interest.

Analysts and observers of special districts typically have sorted themselves into two camps: one that views specialized governance as a flexible, efficient, and responsive institutional design for meeting local service demands, and one that treats special districts as captured by local-growth machines and unaccountable to their constituencies and neighboring governments. The two perspectives make different assumptions about citizens' ability to express policy preferences and about the political incentives for local officials to respond to those preferences. As a consequence, the two camps offer contradictory assessments of institutional performance across a number of different normative criteria. This study offers a conditional theory of specialized governance that reconciles these competing accounts and improves our understanding of the democratic and policy consequences of specialization. By specifying and measuring the effects of special district governance, I also offer new insights about how municipal governments respond to the severity of public problems and the mobilization of local interests.

Local drinking water policy provides the empirical testing ground for this investigation. Management of the nation's drinking water has undergone transformation in recent decades as population growth and environmental regulation have increased competition for access to limited freshwater resources. Drinking water shortages have become a common occurrence even in communities that receive abundant rainfall. In earlier decades, conditions of water scarcity in a region would have prompted the construction of large-scale engineering projects, typically undertaken with substantial state or federal assistance, to expand capacity for water storage, treatment, and distribution. In recent years, however, heightened attention to these projects' environmental and economic costs has reduced their political viability. Taking their place are smaller-scale, decentralized public policies designed to promote water conservation and to distribute existing resources more efficiently and equitably.

Governance has replaced technology in the new era of public water supply management, and local decisions are paramount. Local water systems facing resource constraints must act on their own to reduce water consumption or seek to augment their supply through arrangements with neighboring communities. A local government also might attempt to set limits on a neighboring community's consumption of a shared resource. Meeting future water demand will require difficult policy choices that will favor some water uses over others and will tighten the linkages between land-use planning and water availability. As special districts take on greater responsibility for managing public water systems, it is essential that we understand districts' capacity for engaging in responsive and collaborative decision making in this critical policy area.

The Rise of Specialized Local Governance

Special districts are commonly perceived as shadow governments operating primarily in rural areas, but in reality they are an integral part of local governance in the United States. As defined by the U.S. Census Bureau's *Census of Governments*, special districts are autonomous units with substantial administrative and fiscal independence from general-purpose cities and counties.[1] They can provide almost any of the services of a traditional local government; the main difference is that they perform only a single function or in some cases a few specified functions. Some functions lie outside the scope of specialized governance: special districts do not provide public welfare, and they lack the police and land-use powers held by traditional cities and counties. Although school districts are like special districts in their functional specificity and administrative independence, they usually are treated separately because of school districts' distinct origins, purposes, and domination of local finances. But special districts can provide most local services—including water, sewers, parks, transit, libraries, fire protection, health care, electricity, and airports—and they can range from small, low-budget districts responsible for mosquito abatement to the gigantic Los Angeles

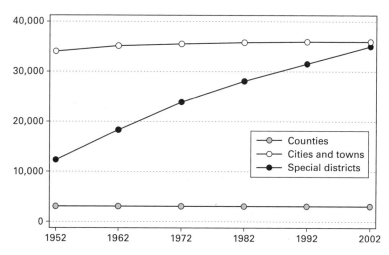

Figure 1.1
U.S. local governments, 1952–2002. *Source:* U.S. Census Bureau 2002.

County Metropolitan Transit Authority and the Port Authority of New York and New Jersey, each with annual expenditures of more than $2 billion.

American communities have become increasingly reliant on special districts over time. As shown in figure 1.1, in 2002 there were 35,052 special districts in the United States, nearly triple the number that existed fifty years earlier. In the same time period, the number of municipal and town governments increased by 6 percent, and the number of counties slightly declined (U.S. Census Bureau 2002b). Special district spending as a proportion of overall local government expenditures also increased during this period, but at a lower rate. In 2002, special districts accounted for 11 percent of local government spending, up from 6 percent in 1952. That amount may seem trivial, but it is important to remember that special districts do not provide education, public welfare, police protection, and corrections—several of the most expensive functions of local government. Setting aside these functions, limited-purpose special districts account for more than 20 percent of local government spending. Their 2002 expenditures on local services nationwide totaled more than $122 billion (U.S. Census Bureau 2005c).

No single function dominates among special districts. One in five districts performs services related to natural resources, but even that category includes a diverse set of tasks, including soil conservation, flood protection, and pest control. Nearly six thousand fire-protection districts make up the second-largest functional category. Special districts are the leading providers of some services: they account for 67 percent of local transit expenditures and 53 percent of local spending on natural resources. They play an important role in a number of other functional areas as well, contributing 40 percent of total local expenditures on housing and community-development services, 37 percent of expenditures on electricity, 36 percent of expenditures on airports and waterports, and 31 percent of expenditures on hospitals.

Although special districts are often treated as rural phenomena, in fact they are more likely than cities and counties to be located within metropolitan areas. Across states, there are no clear patterns in reliance on special districts. Eleven states account for more than 50 percent of all special districts in the United States and 59 percent of special district spending, with Illinois, California, Texas, and Pennsylvania leading in special district expenditures. On a per capita basis, the number of special districts per 100,000 residents ranges from 0.1 in Hawaii to 113 in North Dakota. Just as states vary in their reliance on special districts, they also choose specialized governance for different functions. In most states, for example, operation of public parks lies exclusively within the domain of cities and counties, but 72 percent of Illinois's local spending on parks comes from special park districts. Some states authorize special districts for only a few purposes; others allow communities to set up a district for almost any local government function. More than 80 percent of New York's 1,126 special districts are fire districts; fire districts also make up the plurality of special districts in California, but they account for just 12 percent of all districts.

Further variation exists across and within states on the amount of authority special districts possess. Special districts are independent governmental units, but like all local governments they are creatures of the state.[2] State enabling legislation—either general for a class of districts or restricted to an individual district—specifies districts' functional scope, their authority to levy property or sales tax and collect intergovernmental revenue, their ability to acquire property through eminent domain, and the structure of their governing boards. Because of their diversity in function, jurisdiction, and authority, special districts exhibit even more variation in structural form than traditional general-purpose local governments. This variation provides an excellent opportunity for examining the effects of institutional design.

A mid-twentieth-century observer called special districts "the new dark continent of American politics" (J. Bollens 1957, 1); decades later, many aspects of special districts remain unknown.³ We have a better understanding of the causes of special district governance than of its effects.⁴ Special districts have proliferated because they offer a convenient structure for providing a new public service: they customize service boundaries to the area in need, they allow cities and counties to escape the financial risk of a large infrastructure project, and they satisfy constituents concerned about corruption and mismanagement in existing local governments. Special districts can provide services to specific areas without following the jurisdictional lines of an existing city or county. They can regionalize service delivery to take advantage of economies of scale or localize it to satisfy individual neighborhoods' preferences. The opportunity to create new boundaries can be an advantage when designing the policy response to a problem delineated by natural features, such as a watershed or the habitat of an insect species. It also allows provision of services to new developments that do not incorporate or annex to an existing city.

Special districts are formed in response to local demand for public services, but the actions of other government sectors play a role in creating opportunities for specialized governance. The federal government has provided incentives for special district formation in a number of functional areas, in particular soil conservation and housing during the New Deal and more recently transit and the management of natural resources. But federal policy can also inhibit special district formation: Nancy Burns (1994) has shown that from the 1960s onward, the Voting Rights Act was an obstacle to the establishment of new districts in some counties. State policies are even more important. The most consistent factors contributing to special district formation are the number and breadth of state enabling laws (Burns 1994; Foster 1997). In short, local actors will establish special districts where the state provides the means to do so. State-imposed limits on local general-purpose governments' ability to incur debt or to annex new territory also may contribute to district formation, most likely by reducing the available options for providing services to new development.⁵ Finally, cities may encourage the establishment of special districts in order to meet their own annexation goals (Austin 1998) or to fund projects they cannot afford to administer (Foster 1997; Porter, Lin, and Peiser 1987).

Sometimes it is developers who promote special district formation as an alternative to municipal provision of a facility or service. Special districts' ability to issue revenue bonds, often without any debt limit, allows developers to fund infrastructure for growth without incurring private risk.[6] Kathryn Foster describes the influence of development interests on the establishment of special districts in the latter half of the twentieth century:

As growth controls, environmental regulations, and service moratoria replaced the postwar mentality of growth for growth's sake, property developers found service satisfaction in the relatively autonomous, easy-to-create, politically isolated, financially powerful, and administratively flexible special district. Of particular appeal were districts' bonding powers, which enabled private developers to secure up-front capital for expensive infrastructure projects. Aided often by cooperative public officials and permissive growth policies, developers initiated hundreds of community or subdivision-sized districts to provide water, sewer, drainage, road, street lights, and other development-oriented services. (1997, 19)

Case studies of special district formation offer supporting evidence for developers' influence in creating new districts (Burns 1994).[7] Developers also play a role in city formations, but not to the same extent. City incorporations emerge from a more public and participatory process, and they are more likely to have local residents' active support (Alesina, Baqir, and Hoxby 2004; Burns 1994; G. Miller 1981).

Existing literature offers conflicting hypotheses about the consequences of specialized local governance, predicting that special districts are either more or less likely than cities and counties to be captured by special-interest groups and to deliver inefficient policies that depart from their constituents' preferences. Conventional wisdom treats special districts as invisible and unaccountable to the general public and to their neighboring governments. Critics highlight the lack of transparency in special district operations, arguing that it creates an opportunity for patronage, corruption, and runaway spending. They also charge that political invisibility produces a bias favoring private interests who invest in lobbying special district officials. A *New York Times* editorial expressed the conventional wisdom in characterizing districts as "small, secretive governmental bodies with the powers to tax and collect fees and to hire well-connected cousins, uncles and sons-in-law." It also called them "notoriously costly and inefficient and just as notoriously hard to uproot" ("Mr. Suozzi's" 2007, 15). Another detractor calls special districts "the backdoor government, the invisible government, the shadow government," quagmires of mismanagement and corruption that are unaccountable to the public (Axelrod 1992, 310). Competing with the conventional wisdom is an argument drawn from public choice theory maintaining that specialized governance will enhance public accountability and produce cost savings. According to this view, sorting policy issues into separate, limited-purpose venues provides greater transparency and reduces the costs of communicating with public officials, increasing the likelihood that policy decisions will be efficient and congruent with majority opinion.

The policy effects of functional specialization have received little empirical attention, making it impossible to judge the accuracy of these competing expectations. Some evidence exists to back the claim about the costliness of specialized governance, but assertions about patronage and corruption rest largely on anecdotal support.[8] Critics complain that special districts are difficult to dissolve, whereas supporters of specialized governance applaud special districts for their adaptability to changing problems and conditions (Foster 1997; Frey and Eichenberger 1999; Hooghe and Marks 2003). Most important, little research has examined the representational consequences of specialization or its impacts on public policy outcomes. It remains unknown how specialized governance affects the balance of power among competing interests in a community or the relationships between local officials and their constituents.

Both the conventional wisdom and the public choice framework paint apolitical pictures of special district governance—the former by depicting special districts as operating outside the public's view, the latter by assuming that special districts are purely responsive to constituent demands, efficiently translating those demands into policy outcomes. Neither accounts for diversity across special district functions or variation in district structure and authority. Moreover, both perspectives ignore the political competition that underlies much of local governance. The provision of local services can have important distributional consequences.[9] It also is inseparable from the politics of growth. Just like cities and counties, special districts can be highly politicized arenas for interaction among ambitious officeholders, territorial neighboring governments, resource-seeking bureaucrats, competing interest groups, and attentive neighborhood advocates. The question remains whether an institutional structure that compels specialization influences how conflict among these groups plays out.

This book offers a new theory of specialized governance that is explicitly political. I argue that special district officials are motivated by the same reelection and policy goals as other political actors. The institutional setting affects how these goals translate into policy decisions.

Unlike their counterparts in city and county government, special district officials can dedicate their full attention to a single local function. City and county officials must make trade-offs and agenda choices among a broad range of issues, so their response to a policy question will reflect the policy context—in particular, the severity of the policy problem and its salience relative to other local issues. Special districts' attention to an issue does not hinge on problem severity in the same way, but an issue's salience influences the incentives for interest groups to expand conflicts into special district venues. As a result of these dynamics, the policy effects of specialization are conditional on the public importance of the policy problem. Variation in the institutional form of special districts further influences the policy decisions they make. On the whole, I demonstrate that the effects of specialization are complex and contingent on specific governing structures and on the nature of policies themselves. This contingency makes institutional design a risky endeavor for local actors seeking to create conditions that will favor their policy goals.

The Decentralization of Water Supply Management

Local drinking water policy provides an ideal case for investigating the impacts of specialized governance, in part because of historical factors. Water was the purpose for some of the earliest special district formations; in the late nineteenth and early twentieth centuries, communities established independent districts to regionalize water service and address the growing problem of water pollution.[10] In the West, the success of irrigation districts in securing reliable water supplies for farmers prompted urban communities to consider establishing their own specialized governments for water provision. The popularity of revenue bonds and the imposition of debt ceilings and property tax limitations on general-purpose local governments made special districts an even more attractive option for water governance. Water districts later served as a model for the spread of specialized governance into other local government functions.

Apart from historical context, the study of local water-governing institutions is critically important in the current era because of the rise of the new local politics of water. Increased demand on local drinking water resources has left communities throughout the United States vulnerable to water shortages during periods of drought—as has long been the case in the arid West, where battles over scant water resources underlie much of the region's most contentious politics. Water scarcity is no longer limited to the West, however. Population growth and redistribution have left water systems throughout the country struggling to sate their customers' thirst. When the rain stops falling, newspapers are filled with reports of communities enacting use restrictions and building moratoria in order to stretch out limited water reserves. Seventeen percent of U.S. water utilities responding to a 1999 industry survey reported that they had implemented usage restrictions due to water shortages during the previous five years.[11] A community occasionally runs out of water altogether. In 2002, shortages in the Southeast were so severe that one North Carolina town resorted to importing water by fire hose (Jehl 2002). Drought returned to the region in 2007, requiring the town of Orme, Tennessee, to truck water in daily across state lines. Of course, western states remain most vulnerable. A 2008 study estimated that Lake Mead, the primary water supply for Arizona, Las Vegas, and Southern California, has a 50 percent chance of drying up by 2021 (Barnett and Pierce 2008). Without substantial reductions in demand or new sources of supply, by 2020 California might experience annual water shortages of 2.4 million acre-feet, an amount equivalent to the consumption of five million households (California Department of Water Resources 1998).

This growing struggle to keep pace with local demand for drinking water is a strain on the nation's freshwater resources and has far-reaching environmental and economic consequences. Houston overtapped its groundwater aquifer until the land began to sink, causing property damage and aggravating the region's flooding risk (Perrenod 1984). Groundwater depletion in Tampa has resulted in subsidence, saltwater intrusion, and degradation of local wetlands (Scholz and Stiftel 2005). In Wisconsin, falling groundwater levels have increased the concentration of radium in some communities' drinking water (Gaumnitz, Asplund, and Matthews 2004). Overdrawing from the Ipswich River basin in Massachusetts has reduced surface water flows and caused the river to dry up repeatedly (Glennon 2002). These problems are likely to become more widespread as the escalation of global climate change increases strain on the nation's water resources (National Assessment Synthesis Team 2001).

The historic response to water scarcity was construction of a new dam or aqueduct to increase storage or transport water over a long distance. These large-scale infrastructure projects were typically undertaken with substantial state and federal assistance, and in many cases were led by the U.S. Bureau of Reclamation or the U.S. Army Corps of Engineers.

The grand water projects built during the twentieth century were marvelous feats of engineering that enabled development of the West and many Sunbelt cities. They allowed communities to pursue growth goals with little regard to limits on local water resources. They brought about enormous prosperity and provided to virtually all Americans something that is a luxury throughout much of the world: access to safe, affordable, unlimited drinking water straight from the tap.[12]

These large water projects also carried significant environmental and economic costs, however, and growing recognition of these costs has reduced the political viability of dams and aqueducts as a solution to contemporary water shortages. Heightened environmental regulation rules out many projects that might have been feasible in an earlier era. In some cases, the water simply is no longer available. Claims on the Colorado River exceed the river's flow in most years, and numerous states in other parts of the country are engaged in battles to secure access to rivers that flow across their borders: water-strapped suburbs of Washington, D.C., in Virginia and Maryland are withdrawing all they can from the Potomac; conflict between Virginia and North Carolina over access to the Roanoke River has landed these states in federal court; and Georgia, Alabama, and Florida have been fighting a "water war" for nearly two decades. Even where surplus water might be obtained, it is difficult to win political support for water development. The public no longer backs expensive investments to divert water from its natural course and thus bring about the associated impacts on wildlife, wetlands, and pristine natural areas. New storage projects also attract opposition based on concerns about the possibility that they will stimulate growth. Proposals for major water transfers may falter because of regional loyalties, often incited by resentment over past water projects. And regions such as the Great Lakes Basin that retain plentiful water supplies are acting preemptively to protect their local resources and avoid the risk of future long-distance diversions (Annin 2006).

The aggregate effect of these developments is to hinder construction of new, large-scale water projects. Construction of the Auburn Dam on California's American River halted after a 1975 earthquake, and since then environmentalists seeking to preserve the river canyon have blocked project proponents' repeated efforts to secure funding for the dam's completion. Californians also have consistently rejected proposals for a peripheral canal that would take water from the Sacramento River and carry it around the eastern edge of the San Francisco Bay Delta to pumping plants. Northern Californians charge that the canal is just another water grab by the southern part of the state. A 2003 proposal by developers in Florida for a major transfer from north to south divided that state along similar regional lines. One indicator of the shifting policy environment is that dam removal now receives more attention than dam construction—the number of new dams has dwindled, and scientists and policymakers are beginning to reconsider the value of existing dams and their operation (Doyle, Stanley, Harbor, et al. 2003).

Communities are beginning to develop strategies for managing existing resources more effectively as it becomes more difficult to build their way out of water shortages. Local agencies have limited opportunity to acquire new supply. In addition to getting over the significant hurdles to building storage facilities, agencies must compete with other user groups for access to water resources available locally. Figure 1.2 shows the sources and competing demands for U.S. freshwater resources. Public water supply accounted for just 13 percent of total freshwater with-

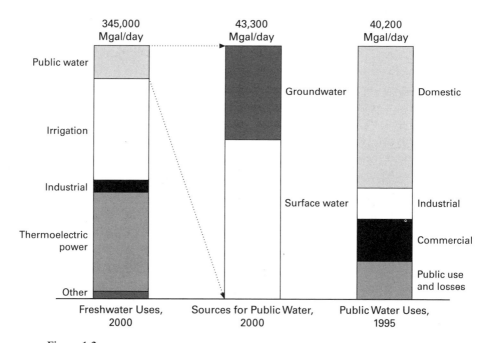

Figure 1.2
Sources and uses of the U.S. public water supply. *Sources:* Hutson, Barber, Kenny, et al. 2004; Solley, Pierce, and Perlman 1998.

drawals in 2000.[13] The biggest consumers of freshwater resources are farmers and power plants; irrigation and the cooling of steam-driven turbine generators account for nearly 80 percent of freshwater use.[14] Setting aside water consumed for cooling in thermoelectric power generation—much of which eventually returns to the surface water body—public water systems' share of freshwater withdrawals totals slightly more than 20 percent. The remaining withdrawals are dedicated to industrial and mining operations, livestock and aquaculture production, and self-supplied domestic consumption.

Water systems' ability to increase freshwater withdrawals is limited by established water rights in addition to overall resource capacity. The majority of the public water supply (63 percent) comes from surface sources, and neither of the two dominant systems for surface water rights favors drinking water uses over any others, nor do they provide clear guidance on allocation during periods of scarcity.[15] The prior appropriation doctrine that is dominant in western states gives priority to senior rights-holders: the maxim "first in time, first in right" demonstrates the importance of long-established claims. Prior appropriation introduces some order to water allocation in times of shortage, reflecting the scarcity conditions that existed in the West at the time of the doctrine's development. Uncertainty remains for holders of junior rights, however, and the doctrine's strict adherence to temporal priority creates a disincentive for cooperative agreements that might help a public water system meet community demands in times of shortage. Senior rights-holders, be they farmers or neighboring public water systems, will be less inclined to conserve and share water resources if they risk forfeiting their right by doing so.

The riparian doctrine prevalent in eastern states provides no greater certainty. It allocates rights based on the land that overlies or adjoins the freshwater source. Because the riparian system assumes abundance, it fails to account for the possibility of water scarcity. Riparian rights have equal priority, so all holders of the rights to a source share the burden of a shortage. In practice, many states are backing away from strict interpretation of either doctrine in order to attach more value to conservation and in-stream water uses as well as to provide clearer guidance for allocation of an increasingly scarce resource (Deason, Schad, and Sherk 2001).

Some states also are beginning to develop systems to regulate extraction of groundwater, at least on a site-specific basis. Rights to groundwater tend to be loosely defined. Like riparian rights, they are tied to the land, but usually without restrictions against storage and transfer to other properties. The lack of regulation over groundwater withdrawals has led to widespread overdrafting; in many places, groundwater pumping currently outpaces recharge. Where aquifers are under particular stress, some states have established groundwater management areas that entail permits and caps on withdrawals.

In sum, a legal framework combines with real resource limits to restrict opportunities for public water systems to seek out new sources of supply. Instead, local utilities must find ways to lower water demand and increase the productivity of existing resources. Much of this activity focuses on domestic users. As shown in figure 1.2, domestic use accounts for 56 percent of the water supplied by public systems.[16] Because domestic users are the largest draw on a water system's resources, reducing demand within that sector can do the most to relieve pressure on a system facing supply shortages.[17] Moreover, consumption is often more discretionary for domestic use than for the commercial and industrial sectors.

The United States has already made progress in reducing the amount of water consumed for irrigation and industrial purposes. Figure 1.3 displays public water as a percentage of freshwater withdrawals from 1950 to 2000. Overall per capita consumption of freshwater resources steadily

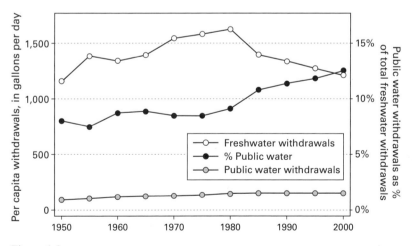

Figure 1.3
U.S. public water withdrawals, 1950–2000. *Source:* Hutson, Barber, Kenny, et al. 2004.

increased after 1950, reaching a peak of 1,625 gallons per person per day in 1980. After 1980, withdrawals for irrigation, thermoelectric power, and industrial purposes declined markedly, thanks to technological improvements and federal regulations that introduced water quality and efficiency standards. These changes have returned per capita withdrawals to their 1950 levels, and the economic productivity of water has improved (Gleick 2003). We have not seen the same conservation gains within the public water sector. Per capita consumption of public-supply resources has steadily increased, keeping approximate pace with the growing number of households that receive public water.[18] With public consumption levels holding steady as other uses become more efficient, public water supply represents a growing percentage of overall freshwater use, rising from 8 percent in 1950 to 13 percent today.

Over time, more people are recognizing water supply as a problem and perceive that they are participating in a solution. Figure 1.4 shows results from a series of nationwide Gallup polls measuring attitudes toward specific environmental problems. In 2000 and 2001, substantially fewer Americans worried about freshwater supply than about pollution of air

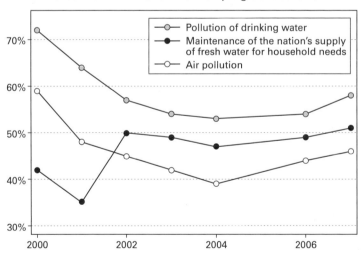

Figure 1.4
Attitudes toward environmental problems, 2000–2007. *Source:* Gallup polls (various).

and drinking water. The percentage of respondents reporting that they worry "a great deal" about having enough water for household needs jumped from 35 percent in 2001 to 50 percent in 2002, and it has remained at approximately the same level in five subsequent nationwide polls. Concern about water supply now scores consistently higher than concern about air pollution.[19] Moreover, people think they are responding to the problem. The percentage reporting that their household tried to use less water over the previous year increased from 56 percent in 1995 to 69 percent in 1999, and then to 83 percent in 2000.[20] Despite these perceptions, real per capita domestic consumption did not decline noticeably over this period. The conservation efforts that people make in their homes get balanced by changes in residential patterns that intensify water demand. Although the average lot size of new homes has declined over time, fewer people live in each household, and the nation's population has redistributed to the warm Sunbelt region, where per capita water consumption is highest.

Only limited demand management can be achieved through voluntary conservation efforts. In the absence of new sources of supply, local water agencies throughout the country are implementing policies that provide stronger incentives for water-use reductions and attempt to distribute existing resources more efficiently and effectively (Beecher 1995). These policies include pricing strategies that send signals about the scarcity of water supplies and the cost of system expansion; contracts and agreements between neighboring governments to share water resources and capital facilities; and procedures for incorporating consideration of water supply into land-use planning decisions. Decision making about water management has become more decentralized as communities and their water providers consider the relative importance of different uses and the appropriate distribution of costs. Meanwhile, the state and federal governments' role has shifted from builder to regulator. Agencies are less likely to help localities build their way out of shortages and more likely to tell them that they must find a way to live within limits.

The devolution of responsibility for water policy can be seen in the management of freshwater resources at their source. Heightened federal regulation over water quality and endangered species has stimulated development of new cooperative institutions for watershed protection and groundwater management. These institutions provide incentives for diverse local actors to negotiate rules for sharing resources and overcoming collective action problems (Blomquist 1992; Heikkila and Gerlak

2005; Lubell, Schneider, Scholz, et al. 2002; E. Ostrom 1990; Sabatier, Focht, Lubell, et al. 2005; Scholz and Stiftel 2005). A decentralized approach to watershed management allows policy solutions that are responsive to problem conditions as well as to local stakeholders' demands and interests. Public water utilities are a key stakeholder in many watersheds, competing for access to freshwater resources with neighboring communities, other user groups, and environmentalists who advocate for increased in-stream flow. In the end, the declining environmental quality of watersheds and the continuing demands upon them likely will compel public water suppliers to withdraw less water than they desire, whether those restrictions are enforced through voluntary partnerships or by command-and-control regulation. This study focuses attention on how local utilities adapt to requirements that they curtail water usage.

Decentralization of water policy also is part of a global trend toward bottom-up strategies for water-resources management. Peter H. Gleick and his colleagues at the Pacific Institute have done extensive work documenting the costs of what they call the "hard path," the centralized engineering approach to water provision (Gleick 2002, 2003; Gleick, Cain, Haasz, et al. 2004; Gleick, Cooley, Katz, et al. 2006).[21] In addition to levying environmental and economic costs in the United States, the hard path has imposed severe social costs in many of the poorest regions on the planet. Dams and reservoirs have displaced populations; river diversions have jeopardized communities' way of life. With a billion people worldwide still lacking access to safe drinking water, the hard path also has failed to achieve its most important goal. International agencies have recently highlighted the drinking water crisis, and the global solutions they propose have much in common with the decentralized approach emerging in American communities (United Nations Conference on Environment and Development 1992; United Nations World Water Assessment Programme 2003; World Water Council 2000).[22] In 2002, the Global Water Partnership declared, "The water crisis is mainly a crisis of governance" (2002, 17).

The United States is one of a handful of countries in the world where access to safe drinking water is universal and largely affordable (World Health Organization 2000). Without question, the stakes for water management are lower here than in nations grappling with widespread waterborne disease. But the decentralization of water management intersects with another kind of crisis looming for American water systems: a financial crisis brought on by the deterioration of the nation's water infrastructure. The U.S. Environmental Protection Agency (EPA) predicts that public water systems will need to invest $270 billion over the next 20 years in order to replace deteriorating storage, treatment, and distribution infrastructure and to ensure compliance with federal water quality regulations (U.S. EPA 2005). Existing revenue from water sales and from state and federal assistance programs falls far short of that sum, and the agency has warned about the possibility of a significant funding gap between needs and spending (U.S. EPA 2002b). In proposing strategies for closing the gap, the EPA has offered the concept of "sustainable infrastructure," which includes increasing water efficiency, implementing full-cost pricing, and other policies that are consistent with the new, decentralized approach to water management. The decentralization that is helping communities address water shortages and manage their resources more effectively in the short term may also help in the long term to limit the financial damage caused by more stringent environmental regulation and the aging of the nation's infrastructure.

Specialized Governance and the New Local Politics of Water

Responsible water planning in the current era requires a clear understanding of local conditions related to water quality and supply, coordination across jurisdictional boundaries, and responsiveness to community preferences. It demands effective and accountable governance.[23] In attempting to address water supply challenges, communities are adopting policies that redistribute costs between existing and future residents and impose private costs in order to achieve public benefits. Decisions to extend water lines have consequences that spill over geographic and functional boundaries. The allocation of a scarce resource is an inherently political question. It has important regional consequences as well, because a community's water policies help determine future development patterns.

Debate over privatization has dominated conversation about local water-system management since Indianapolis, Milwaukee, and many smaller cities and towns began contracting with private companies in order to reduce the costs of providing drinking water.[24] In some cases, privatization proposals have sparked public debates about private firms' accountability to local residents' interests and about their ability to protect water quality and affordability (Jehl 2003; Reiterman 2006). Atlanta and the city of Stockton, California, ultimately took back control of their water systems from private contractors, the latter after the contract failed to survive a court challenge brought by citizens' groups. Notwithstanding these high-profile controversies, private firms in fact tend to operate only the smallest water systems: slightly more than half of community water systems are privately owned, but they produce only 9 percent of the total public water supply (U.S. EPA 2002c).[25] The great majority of Americans receive their drinking water from a utility that is operated by a local government.

Overlooked in the debate about water privatization is the rise of specialized governance among publicly owned water systems. Between 1962 and 2002, the number of special districts involved in water supply nearly tripled.[26] Water districts now account for 28 percent of local government expenditures on water supply. Specialized public governance receives less attention than privatization from industry analysts and the public, yet empirically it is more common. Such inattention is surprising, considering that much of the debate over privatization focuses on private water firms' accountability and responsiveness—the same issues raised by critics of special districts.

Given the rising importance of local decision making in addressing the nation's water supply issues, we must consider whether special districts are up to the task. If special districts are biased institutions as the conventional wisdom suggests, they may be less likely to pursue a public good such as water conservation if the costs fall on influential special interests. Specialized governance also may interfere with the cooperation needed to address local water supply challenges. Efficient distribution of water resources will sometimes require contracts and agreements between neighboring jurisdictions for cost sharing or the transfer of resources. It also will involve greater coordination between water and land-use planning. Water and land use are inseparable—new development requires a reliable water supply, and patterns of land use lock in water demand and groundwater replenishment for the long term. Yet planning processes historically have ignored these interrelationships. With the growing scarcity of water resources, communities are beginning to integrate planning for water and land use, sometimes under pressure from state government.[27] As Atlanta's commissioner of watershed management described the change, "This city had a motto for years, and it went something like 'Atlanta grows where water goes.' I think we've learned enough to know that we'd prefer to see the city in charge of that destiny" (Jehl 2003, A1). Coordinating water and land use may be a more profound challenge when a specialized water district governs the tap.

Book Overview

This book offers a new theory about the policy effects of specialized governance and tests that theory in the domain of local drinking water management. At its heart is a series of empirical analyses that directly compare the policies enacted by water districts with those created by cities and counties that operate their own water utilities. These policies provide insight about the existence of bias in governmental responsiveness and the possibility of intergovernmental coordination between special districts and their neighbors. In addition, the book investigates the broad range of special district structural forms and demonstrates that rules governing elections and boundary change further shape incentives for special districts to respond to their constituents and cooperate with neighbors.

Local public water utilities are part of a complex institutional network consisting of wholesalers and retailers, state regulatory agencies, regional bodies designated to protect sources and watersheds, and cooperatives and water districts established to provide irrigation water to agricultural users. This study sets aside most of this network in order to concentrate on the retail provision of drinking water, primarily for household use. Water is a natural monopoly, so residents and businesses in a given location rarely have a choice among providers and cannot exit service without physically relocating.[28] This feature allows me to assume that all residents of a given water jurisdiction are affected by the water utility's policies and thus allows direct comparison across governance types.

The analysis is national in scope. Water districts exist in almost every state, but in no state do they have universal control over retail water provision. Figure 1.5 shows special district spending as a percentage of total local government spending on water supply at the state level. It ranges from 0 to 91 percent across the 50 states, with a mean of 22 percent of a state's local water spending being allocated by water districts (U.S. Census Bureau 2005c). The number of water districts in each state appears in figure 1.6. All states except Alaska and Hawaii have at least one independent water district.[29] In some communities, drinking water is part of the package of local services overseen by elected city or county officials, but in other communities residents receive a water bill from a specialized government responsible only for water provision. This study assesses the policy effects of that variation.

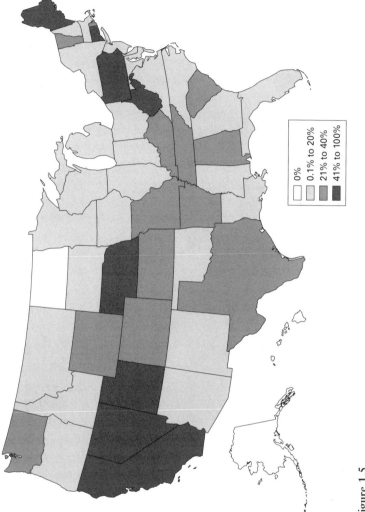

Figure 1.5
Special districts' share of local spending on water by state, 2002. *Source:* U.S. Census Bureau 2005.

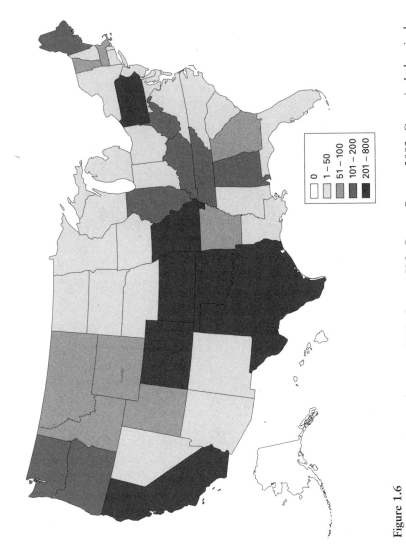

Figure 1.6
Number of water districts by state, 2002. *Source:* U.S. Census Bureau 2002. Count includes single-function districts focusing on water supply and multifunction districts providing water supply and sewerage or water supply and natural resources.

Chapter 2 presents the conditional theory of specialized governance that guides the empirical analyses. The theory reconciles the conventional wisdom treating special districts as invisible and unaccountable to the public with public choice accounts that predict greater policy responsiveness in a system that is fragmented along functional lines. I argue that both of these frameworks oversimplify the dynamics of special district governance by assuming constant effects of specialization across issues and political contexts. The conditional theory takes seriously both the function that a local government performs and features of a special district's institutional design. It predicts that these factors will condition the impact of specialized governance on policy outcomes.

Chapters 3 through 6 discuss and examine a number of recent policy innovations in water planning and test the effects of specialized governance on policy adoption. The first three of these chapters rely on quantitative data from national and state surveys of public water utilities. The main text emphasizes the substantive meaning of the findings, with fuller detail on methods and results appearing in the appendixes. Chapter 3 examines adoption of progressive rate structures that offer the promise of economic efficiency, water conservation, and income redistribution while imposing concentrated costs on the wealthiest members of a community. The analysis demonstrates the impact of institutional design on how a government balances public goods and private demands. Chapter 4 investigates another water-pricing strategy, the use of development impact fees to fund the cost of water-system expansion. Water systems must weigh constituent demands to pass on the costs of growth to incoming residents against developers' opposition to these fees.

In chapter 5, the focus shifts from bias in policy outcomes to patterns of intergovernmental cooperation. It explores the flexibility of special district boundaries and evaluates the relationship between boundary flexibility and establishment of interlocal agreements that might promote efficiency and equity in water management. Chapter 6 investigates interest-group strategies and intergovernmental coordination in a series of local growth disputes in California and Pennsylvania. The chapter draws on interviews, lawsuit briefs, and other qualitative data to evaluate how separating responsibility for water and land use influences the politics of growth. The final chapter reviews the book's main findings and discusses how specialized governance might affect local capacity to promote sustainability and confront the challenges presented by global climate change.

2
A Conditional Theory of Specialized Governance

Political scientists who study urban politics have long recognized that institutional design helps determine who benefits from decisions made by local government. Focusing on the reform structures adopted by cities early in the twentieth century, scholars have argued that a city's form of government and its rules for elections affect different groups' relative influence in a community. As a result, the organization of municipal government can shape patterns of representation and may have an impact on policy outcomes.[1]

Functional specialization is a more striking institutional departure. Specialization allows decision makers to develop issue expertise and reduces logrolling that may heighten efficiency but can distort policy decisions away from outcomes the majority would prefer. Specialized governance also increases the number of policy venues, reducing the visibility of each individual venue. Crosscutting jurisdictional boundaries may carve up communities of interest and rupture channels of communication between neighborhoods and public officials. When these competing dynamics are taken into account, it is difficult to predict how special district governance will affect policy outcomes. Existing literature offers competing hypotheses about the policy effects of specialized governance but provides little systematic evidence to arbitrate between them.

Previous theorizing has assumed that special district governance has the same impacts regardless of the district's institutional design, the political environment, or the nature of the district's public function. Perhaps that explains the dissonance in predictions about specialization's consequences. Many factors influence how local officials make policy decisions, including constituents and interest groups' demands, constraints on local autonomy, and the severity of the public problem. Elements of the policy context influence how these forces interact. Different problems place different demands on public officials. In order to understand how specialization influences policymaking, we must consider the multiple factors that shape policy outcomes in traditional city and county venues.

In this chapter, I present a conditional theory of specialized governance that accounts for the diversity in local policymaking contexts. The theory combines elements from previous models of special district governance and describes the circumstances that influence those models' explanatory power. The conditional theory proposes that specialization has effects that vary in complex but predictable ways. Depending on problem conditions, specialization may create more responsive policymaking and at the same time pose greater challenges for cooperation across jurisdictional boundaries. An overall evaluation of special district governance, therefore, must take seriously both the context in which it exists and the normative values we seek to maximize. It also must address the variable performance of general-purpose governments across problem contexts.

Independent, Overlapping, Special-Purpose Governments

Three features distinguish the special districts under study in this volume from other types of local public entities: independence, territorial overlap, and limited functional scope. My interest lies only with independent special districts, not with dependent districts or quasi-public boards and commissions that existing governments establish and operate to build and manage public works. Dependent districts help cities circumvent ceilings on debt and taxation as well as other administrative requirements such as contracting rules and civil service regulations. They isolate debt and revenue for a function from the rest of the municipal budget, thus reducing pressure to cross-subsidize between the function and other city services. Dependent districts make it easier for cities to provide public goods that local constituencies demand, but they have been attacked for being unaccountable to the communities they serve and for bolstering efforts by state and suburban leaders to gain control over infrastructure in center cities (Adams 2007; Axelrod 1992; Henriques 1986; Walsh 1978). Most important for my purposes, dependent districts do not make local governance more specialized, because ultimate authority over their activities lies with the cities and counties that establish them.

Many scholars have drawn a distinction between independent special districts and dependent public authorities based on an entity's taxing authority and its process for selecting board members (Adams 2007; Doig 1983; Eger 2006; Mitchell 1990; Walsh 1978). These definitions typically treat special districts as independent only if their boards are elected rather than appointed and they have the power to levy taxes. In practice, the definitions have little correspondence to the labels used in state statutes, and local entities established by those statutes may have a diverse mix of rules for board selection and revenue collection (Leigland 1994). An entity with the same power and governing structure can be called a "special district" in one state but a "public authority" in another. In California, an independent special district may be categorized as enterprise or nonenterprise—either funded entirely through user fees or reliant on property taxes for revenue—and its board may be elected or appointed for fixed terms. Imposing definitions that do not match practical usage only confuses discussion about a class of governing institutions that already are unfamiliar to most of the public.

Moreover, neither board selection nor taxing authority reveals a special district's actual level of autonomy from other governments. If the district has its own governing board with members appointed for fixed terms and if it exercises sole authority over policies that distribute the costs and benefits of a public service, it is difficult to see how that district is dependent on any other government. For example, the municipal authorities that proliferate in Pennsylvania lack taxing authority and have governing boards appointed by officials of the overlapping city or township. After appointing an authority's board, however, a municipality has little influence over the authority's operation. Board members serve overlapping five-year terms. Once established, authorities are independent agencies, and municipal officials in Pennsylvania who want to influence an authority's decision making must wait for opportunities to appoint sympathetic district officials (Governor's Center for Local Government Services 2002).

This study is concerned with all entities that have fiscal and administrative independence from other governments, recognizing that these autonomous entities may operate under the title "special district," "public authority," or some other variation. To identify independence, I follow the U.S. Census Bureau's definition, which disqualifies districts that are unable to set their own budget or issue debt without approval by another local government. Districts must perform their own functions without oversight by any other government entity. Census-designated independent districts may have elected or appointed boards, but their board cannot be composed entirely or mainly of officials from the government that created the district (U.S. Census Bureau 1999, x).[2] For the purposes of my analysis, there is little difference between a water utility that is a department of city or county government and a water district that is overseen by the city or county council temporarily convening as a special district governing board. In both cases, city or county officials are responsible for water policy. The latter arrangement may provide more administrative flexibility, but it will not change the political constraints and incentives that guide local officials' behavior. This study treats only districts with their own boards of directors, either elected or appointed by officials of another government, as special districts. Taxing authority is not necessary to qualify a district as fiscally independent; the district need only the authority to make its own fiscal decisions within the taxing and spending constraints defined in its enabling statute.[3]

The second important feature of special districts is their geographic flexibility, resulting in district boundaries that intersect one another as well as the boundaries of neighboring cities and counties. Establishing a government to perform one rather than many functions allows design of territorial boundaries according to the requirements of that function. General-purpose governments do not have the same flexibility, in part because of the demands of governing across multiple policy areas, but more important because their boundaries may not overlap. Cities and counties are hemmed in by existing boundaries, but special districts can be superimposed on top of these historical divisions. In 2002, just 16 percent of water districts and 30 percent of special districts overall had boundaries that corresponded with those of a city or county.[4] Most district boundaries do not conform to existing jurisdictions, but instead address problem-specific demand—or reflect a political compromise reached at the time of district formation.[5] In regions with high levels of special district reliance, the dense networks of intersecting boundaries can result in a multitude of configurations for service provision. Neighboring households may share a water provider but fall into different fire districts; they might be taxed to support the same parks but have library privileges in separate jurisdictions. For some services such as public transit, overlapping jurisdictional boundaries can produce competition among service providers for a household's patronage. More commonly, a government has a territorial monopoly to provide a service but competes with other functions and jurisdictions for the household's revenue dollars (C. Berry 2007).

Finally, what separates a special district from a city or county is its limited purpose. State statute defines a special district's functional scope, either through a general act that enables a class of districts or specific legislation that creates an individual entity. Either way, state law establishes the breadth of district authority and typically restricts it to a single function. In 2002, more than 90 percent of special districts performed only one function. Combined water-sewer districts account for almost half the multifunction districts (U.S. Census Bureau 2002b). Some states enable formation of broad community-services districts that can provide multiple public services in an unincorporated area, but these entities make up a small minority of the overall special district population.

The combination of these three features—independence, territorial overlap, and limited purpose—defines special districts and distinguishes them from other types of public organization. In contrast, other independent local governments perform a wide array of functions and have boundaries that are largely fixed. Bureaucracies and dependent districts have limited purpose and may have geographic flexibility, but they are not independent. Regulatory agencies that have formal structures designed to insulate them from political influence still lack autonomy over their budgets and the judicial actions they bring, and their geographic boundaries coincide with those of the establishing government (Moe 1982).

Special districts' institutional design promotes development of issue expertise while maintaining the political incentives that operate for other independent governments. Early students of the bureaucracy, including Max Weber (Gerth and Mills 1946) and Woodrow Wilson (1887), proposed models of administrative neutrality in which elected representatives would set public policy and appointed, neutral issue experts would implement it. The idea that neutrality is a feasible or even desirable goal has long since come under question, but scholars remain concerned with problems that arise when elected officials delegate policy responsibility to bureaucrats.[6] State and local officials involved in forming a new special district may perceive that they are delegating authority to a subordinate institution, but there is no guarantee that an independent special district will respond to the interests of the actors involved in its establishment.

Those who turn responsibility for a function over to a special district also might imagine that they are liberating the function from political influence by putting it in the hands of technically competent, neutral administrators. This assumption is equally likely to prove false. Service-delivery decisions have important political consequences, and powerful interests in a community attempt to influence those decisions regardless of where they take place. Although special districts may be less visible than cities and counties, interested citizens and organized groups will use lobbying and electoral strategies in order to pursue their goals, just as they do in traditional public venues. As the Advisory Commission on Intergovernmental Relations (ACIR) observed several decades ago, "political decisions remain political decisions whether made by a unit of general government or a special district" (1964, 53).[7] Specialized governance might alter the political incentives that influence local policymaking, but it does not eliminate politics altogether. When a policy decision involves the distribution of scarce resources, politics are inescapable.

Existing Theories of Specialized Governance

Much of the theoretical attention special districts have received arises from a larger debate over the ideal form of metropolitan organization. On one side of the debate are the polycentrists who favor the fragmentation produced by multiple competing governments operating within a region. On the other side are the supporters of the reform tradition in local governance who prefer a consolidated body to govern the region as a whole. The debate has focused primarily on geographic fragmentation, or the density of municipalities in an area, but scholars on both sides extend their arguments to the functional division of responsibility among limited-purpose governments. The two perspectives offer different accounts of the incentives that specialization and fragmentation create for local political actors, and they produce competing predictions about the consequences of specialized governance for efficiency, responsiveness, and policy coordination within and between communities.

Metropolitan Reform Theory
The dominant view of special districts comes from reformers who argue that specialized governance creates redundancy, waste, and an opportunity for progrowth special interests to dominate local policymaking. Rooted in the Progressive Era reform tradition, this framework treats metropolitan organization as a management problem. The unit of analysis is the metropolitan region, and reformers ask how political systems should be organized to maximize benefits for the region as a whole. They conclude that the best solution is to consolidate governing struc-

tures across geographic and functional boundaries in order to create large, multifunction governments that integrate management of a region's public services.[8]

Metropolitan reform theory evaluates efficiency according to the costs of service provision and predicts that overlapping jurisdictions will result in wasteful and costly service duplication. Conflicts between governments whose boundaries intersect make it more difficult to negotiate and maintain agreements about service responsibility. Moreover, the fragmentation of authority among governmental units reduces the visibility of each individual unit, hampering the public's efforts to hold special districts accountable for their policy decisions. As John Bollens described in the leading reformist study of special districts in the late 1950s, "Special districts have multiplied so rapidly that citizens no longer keep themselves well informed on this aspect of governmental affairs.... Although conscientious citizens might conceivably have exercised effective control over a few governmental units, it was unreasonable to expect them to watch and regulate a multi-ring circus" (1957, 252–253). Reformers predict that this lack of accountability will give rise to runaway spending, further increasing the costs of service delivery.

Adding to the problems caused by fragmentation of authority are institutional features that depart from the democratic procedures and accountability mechanisms we expect in local government. In many special districts, governing officials are selected by appointment rather than by election. Where elections do occur, they often take place off the regular election cycle, reducing voter participation.[9] In some cases, special district elections violate the principle of "one person, one vote." The Supreme Court has upheld franchise limitations for special districts, opening the way for districts to confer voting rights based on property ownership.[10] Landowners need not reside in the district to vote, and some districts even apportion voting power based on amount or value of land held. Appointment of district officials and property-based voting rules add another obstacle to the public oversight of special districts that reformers view as essential for controlling government spending.

According to reform theory, the lack of public scrutiny over special districts not only contributes to wasteful spending, but also reduces the responsiveness of policymaking to constituents' interests.[11] Reformers argue that districts' low public profile, complex governing structures, and jurisdictional boundaries that divide communities of interest serve to confuse citizens and keep them from expressing preferences and grievances to their representatives. Where district officials are elected by popular majority, residents lack the time and information they need to cast informed votes; where officials are appointed or elected through exclusionary processes, residents may not even have the opportunity to participate. In the absence of public oversight and electoral accountability, reformers predict that special districts will develop a bias that favors groups with a concentrated interest in specific functional areas. Special-interest groups are the only ones with the time and resources necessary to monitor district activities and lobby district officials. Victor Jones, an early reform scholar, argued that a "separate government for each function would be the ideal solution of the problem of governmental areas from the point of view of single-interest groups" (1942, xxi). Specialized governance allows these groups to dominate policymaking within their area of interest and to avoid participating in city or county government. Their absence from the broader venues of local politics, Jones argued, "weakens the general government for its most important function of bringing the complementary and divergent interests of a locality together into a community" (1966, 240).

To reformers concerned about integrating policymaking within metropolitan areas, perhaps the biggest drawback of specialized governance is its potential for creating coordination problems.[12] The provision of a public good or local service can produce externalities that spill over territorial and functional boundaries. For example, a positive geographic spillover occurs when a city establishes a public park that residents of neighboring communities can enjoy; a functional spillover exists if a water district's upgrade of its distribution system results in improved water pressure for firefighting. Policy spillovers are frequently negative, however, and they can be an important source of conflict between neighboring or overlapping governments. Many negative spillovers relate to growth and land use. The extension of infrastructure into previously undeveloped areas affects environmental quality and creates new demand for public services, and it can induce unwanted growth in nearby communities. Reformers predict that specialized governance increases the incidence of negative externalities because of fragmentation of policy authority, redundancies in service provision, and districts' bias toward progrowth special interests. Consolidation would create fewer opportunities for jurisdictions to adopt policies that negatively affect their neighbors and would provide a centralized venue for negotiating trade-offs across functional areas.

Through its influence on journalists and good-government groups, reform theory has come to dominate public perceptions about specialized governance. In states that are reliant on special districts, support builds periodically for measures to halt the proliferation of districts and to consolidate districts that already exist.[13] These efforts occasionally bring about new accountability measures or procedural hurdles to district formation, but they are rarely successful in producing fundamental reform of local government. Proponents of the "new regionalism" recently have laid new charges against metropolitan fragmentation, contending that it contributes to racial and economic polarization (Orfield 1997; Rusk 1993) and fails to address environmental and social challenges that are fundamentally regional in scope (Downs 1994; Katz 2000). Unlike the midcentury reform literature, however, this new metropolitan reform theory is ambivalent about special districts. Although some see districts as an obstacle to regional coordination (Hamilton 2000), others view regional special districts as a politically feasible step toward general-purpose metropolitan governance (Altshuler, Morrill, Wolman, et al. 1999; Downs 1994; Pagano 1999).

Public Choice Theory

Metropolitan reform theory begins by focusing on the metropolitan region and asks how to coordinate governance at the regional level. In contrast, public choice theory begins with the individual, assumes that he or she will behave rationally, and examines how different governing structures create incentives that guide individual behavior and influence decisions about policy and service delivery. Scholars working in the public choice tradition reach the opposite conclusion from reformers: they contend that local governance will be most efficient and responsive if authority is fragmented among overlapping units whose size and functional scope are designed to meet the demands of the activities they perform.[14]

Public choice scholarship embraces special districts for creating competition among local governments, which provides an incentive for each government to operate more efficiently. Drawing on Charles Tiebout's (1956) model of local governance as a marketplace in which jurisdictions compete for residents by offering rival packages of public goods, public choice scholars argue that functional specialization promotes economies of scale and makes available a wider range of tax and service bundles than would be available from a single general-purpose government, thus allowing a better match between residents' preferences and public services. Dividing responsibility for public functions is expected to contribute further to a competitive service economy by allowing appropriate scaling of public goods, separation of the production and the provision of services, and flexible contracting arrangements among local governments (Ostrom, Tiebout, and Warren 1961).[15]

Public choice theory measures efficiency not strictly by the cost of providing services, but rather by the quality of the match between residents' preferences and service levels. An efficient local economy in this framework minimizes waste by allocating resources to maximize the benefits from their use. High spending on a service does not indicate inefficiency if local residents prefer a high level of service. Because smaller government units reduce the diversity of opinion within each unit, they improve the match between individuals' preferences and service levels. Disentangling functions from one another through specialized governance also makes it easier for residents to express their preferences about a specific service, thereby increasing both efficiency and responsiveness in service provision.

The vote enhances the performance of fragmented systems by offering a tool for residents to express policy preferences and hold local officials accountable. Elections provide an opportunity to evaluate incumbents' performance and prospective candidates' policy positions. Citizens can use their vote to signal demand for more or less spending or to register an opinion about a salient policy conflict. A vote sends a clearer message about policy preferences when the offices being contested oversee a limited range of functions:

> Voters often face a "blue plate" menu problem, in being forced to allow a single action—a single vote—to express preferences on many issues.... A priori, the choice for delegates will be more efficient if the delegate is responsible for a small related set of public activities rather than for a large number of unrelated public activities because in the more limited area, the individual voter has more chance of finding a delegate whose positions coincide with his own. (Bish 1971, 70)

Multiple and overlapping service providers allow expression of more complex sets of preferences because residents can separate their preferences on parks from their preferences on public safety or land use and can evaluate public officials for their positions on the specific functions they oversee. Thus, the vote becomes a more effective mechanism for ensuring accountability and responsiveness. In a multidimensional election, officials cannot discern what issues drive the vote calculation. In a special district election, votes more clearly express preferences about specific policy functions. Special districts also allow more flexibility for meeting demands for a public good that deviate from the average (Hawkins 1976; Ostrom, Bish, and Ostrom 1988).

Another way for residents to express issue preferences is to communicate directly with local officials, and public choice theory contends that such communication also becomes easier when governing structures are organized along functional lines. The layering of independent governments provides citizens with multiple access points, reducing the cost of communicating with government officials. Public officials will face fewer demands on their time due to the smaller portfolio of functions they oversee. When a single government is responsible for all services, it can be difficult for residents to gain access to public officials; as a consequence, politicians will be most likely to hear from those residents who have the most political resources. Organized lobbying may be necessary to capture the attention of public officials in general-purpose governments, creating a bias in favor of interests concentrated on a particular policy issue and with resources and experience in communicating with government officials.

Finally, public choice scholarship offers an optimistic assessment of interlocal coordination in regions with fragmented governance. In the public choice framework, one key to achieving efficiencies is the development of contracting networks that allow the production and provision of public goods each to occur within boundaries that take advantage of scale economies, minimize negative spillovers, and ensure political representation. The existence of multiple producers and providers promotes healthy competition among governments as well as the development of cooperative relationships that provide communities with even more flexibility to meet service demands. Some public choice scholars acknowledge that there may be contexts in which multiple producers lead to a net loss in efficiency, but in general they predict that the costs of reaching and enforcing agreements will be low enough not to cancel out the benefits of fragmentation (see E. Ostrom 1972).

Empirical Literature

The debate over metropolitan organization has produced a sizeable empirical literature testing the microfoundations of Tiebout's model and the efficiency effects of fragmentation among general-purpose governments.[16] Special districts receive peripheral attention at most in these empirical tests. Despite considerable theoretical development, we have little empirical evidence on the policy effects of functional specialization. The few studies that directly compare spending of special districts and general-purpose governments produce competing results, finding that specialization either raises or lowers spending on a function (Mehay 1984; Minge 1976). Aggregate analyses of the effects of district proliferation suggest that district reliance tends to increase local spending (DiLorenzo 1981; Foster 1997; MacManus 1981). This result does not necessarily refute public choice assertions about special district efficiency, however, because district reliance may be a sign that local residents prefer higher levels of service.

No previous study has examined responsiveness and policy coordination across a large number of special districts. However, case study evidence lends support to the criticisms lodged by metropolitan reformers. A number of studies have argued that the invisibility of special districts allows them to be captured by developers pursuing a progrowth policy agenda. Reviewing the results, Kathryn Foster concludes, "The most fundamental finding is that special districts influence public policy in ways often inconsistent with public goals, particularly with respect to growth and development. Although some studies... document districts' ability to slow growth, most underscore the... contention that districts foster prodevelopment agendas" (1997, 77–78). Notable examples of developer-dominated districts include the municipal utility districts (MUDs) that developers set up to provide water and sewer services to new housing near Houston (Perrenod 1984; Porter, Lin, and Peiser 1987; Thomas and Murray 1991) and water districts serving the vast arid urban landscape of southern California (Gottlieb and FitzSimmons 1991). Texas MUDs historically have served as an instrument for development, allowing infrastructure expansion to keep pace with the state's rapid growth. In the absence of public accountability, many utility districts engaged in financial misdeeds and failed to maintain environmental standards, ultimately leading to groundwater depletion and land subsidence. Even after the state increased oversight of MUDs in the 1970s, developers continued to use the districts as a tool to achieve their growth goals (Egerton and Dunklin 2001; Thomas and Murray 1991). Southern California water districts receive more careful public scrutiny than the Texas MUD received, but some analysts have argued that the districts' strong preoccupation with promoting growth has imposed environmental and social costs on the region.[17]

These case studies indicate that special districts can become dominated by developer interests, but it is not clear that limited-purpose governments are any more progrowth than the cities and counties they overlap. Researchers who have examined Houston's MUDs and southern California's water districts have not explored the counterfactual scenario of whether policy outcomes would be any different if a general-purpose government had made water decisions. The invisibility of water agencies in both cases may facilitate decision making that favors developers, but there is little reason to expect a different outcome had a city or county been in charge. In the case of southern California, Los Angeles County is notorious for its progrowth outlook and expansionist history. The Department of Water and Power for the City of Los Angeles was instrumental in obtaining the water that allowed the city to boom (V. Ostrom 1953); it was surrounding communities' concerns about Los Angeles's monopolistic use of water for expansion that prompted formation of the region's Metropolitan Water District (Gottlieb and FitzSimmons 1991). In Texas, Harris County and the City of Houston similarly share a progrowth outlook, and MUDs are more an instrument of the city than an alternative to it. Utility districts allow development without the city's or county's taking on new fiscal obligations; by the time incoming residents have paid off some of the debt incurred for the development, Houston typically has annexed the area. The arrangements create a partnership between developers and the city, which together make decisions about development without participation from any other actors (Thomas, Hawes, and Calderon 2003). Oversight of water service by the City of Houston rather than by a patchwork of MUDs might have reduced the pace of development there, but for reasons of institutional capacity rather than because of policy choice.

In sum, then, the empirical evidence comes up short in measuring the impacts of specialization on efficiency, responsiveness, and policy coordination. Although results are mixed, the balance of evidence suggests that special district governance tends to increase the cost of local public services. But it is not clear whether the costs result from redundancy and wasteful spending, as reformers would expect, or from higher service demands from special district constituents. Moreover, we have little evidence that indicates whether policy responsiveness and intergovernmental cooperation are possible within a system of specialized governance. This volume aims to address these questions.

A Conditional Theory of Specialized Governance

Both dominant theories of specialized governance assume that specialization will have the same effects across functions and policy contexts.[18] Each theory points to a set of institutional structures and attributes differences in policy outcomes to the presence or absence of those structures. For metropolitan reformers, the primary offending institutions are the selection processes for special district officials. Selection by appointment rather than by election, property-based voting rules, and off-cycle elections, they argue, reduce the visibility of district officials and their accountability to voters. Absent strong reelection incentives, district officials become less responsive to their constituents. Fragmentation of authority makes it difficult for citizens to monitor each individual government and increases the costs of policy coordination on complex, regional issues. In a system of limited public oversight and strong incumbent protection, reformers argue that district officials can deliver policy outcomes that favor developers and other private interests without worrying about staying in office.

Public choice theory finds the source of bias in multidimensional elections. In the classic median-voter model of majority elections, voters' policy preferences align along a single dimension.[19] Politicians then craft policies that appeal to the pivotal voter, whose preferences dictate election and public policy outcomes. In multidimensional elections, however, an individual's vote does not make a clear statement about policy preferences, especially in local elections, where issues often have little ideological content. A city council member who wins reelection over a well-funded challenger might not know whether voters expressed approval of her leadership in firing the police chief or of her support for economic development. With low rates of turnout and noisy signals from election results, direct appeals become more important in guiding local politicians' decisions. Constituents do not have equal ability to make effective demands, however, and those who are most articulate and effective in organizing on their own behalf will have the most influence in a general-purpose venue (Ostrom, Bish, and Ostrom 1988). In contrast, the vote in a specialized election contains more information, reducing the influence of direct lobbying on policy decisions. The outcome, public choice scholars argue, is more equal representation in specialized venues and greater responsiveness to the median constituent.

Both of these theories oversimplify the complex and variable dynamics of policymaking in both specialized and general-purpose venues. The property-based voting rules that reformers highlight merit attention from the standpoint of legal principles, but empirically they are uncommon.[20] Appointment of district officials is more widespread, but the majority of district officials win office through popular election where voting rights are shared equally. Although special districts often operate outside the public's view, in most cases voters have the opportunity to replace district leadership if policies depart too much from public preferences. Moreover, even where districts are more costly to monitor, active neighborhood and interest-group organizations can counter developers' efforts to dominate district decision making.[21] Just as reformers overlook the many districts with effective accountability mechanisms, public choice scholars mistakenly assume that these mechanisms exist for all districts. Instead, there is important variation across districts in their structural organization—for example, in the location and flexibility of boundaries and in the opportunities for electoral participation. These institutional factors most likely have an impact on special districts' performance in responding to constituents and cooperating with neighbors.

Predictions about the impact of specialization also fail to account for variation in the behavior of multipurpose legislatures across policy issues. In a general-purpose legislature, decision makers must be selective about where they allocate attention (Jones and Baumgartner 2005). The breadth, volume, and complexity of their workload make it impossible to engage in fully rational decision making on every policy issue. Legislators use informational shortcuts to reach decisions on some policy questions. They delegate other questions to bureaucracies or leave them off the agenda altogether. In short, decision processes vary across issues in a multidimensional legislature, with some issues taking priority over others. Thus, the effect of establishing a specialized venue for decision making should be conditional not only on the institutional structure of special districts, but also on the attention an issue receives in a traditional legislative venue. This conclusion is the core of the conditional theory of specialized governance I propose and depict in table 2.1.

Although the conditional theory that I propose departs from the dominant treatments of special district governance, it builds on a long line of urban politics scholarship emphasizing the mediating role of institutions. A government's institutional organization can affect its responsiveness to

Table 2.1
The Conditional Theory of Specialized Governance

Conditioning variables	Predicted effects	
	Responsiveness	Intergovernmental coordination
Problem severity	↑ responsiveness of cities and counties ↓ policymaking activities by special districts outside districts functional boundaries	↑ benefits of cooperation
Special district elections	↑ responsiveness of special districts	↑ scope of conflict on complex issues ↑ benefits and costs of cooperation
Special district boundary flexibility		↓ benefits of cooperation
Contiguity between special district and city or county boundaries		↓ scope of conflict on complex issues ↓ costs and ↑ benefits of cooperation

different demands and its capacity to address policy challenges. Structural factors can promote or stifle electoral competition and policy change, and they can stack the deck to favor certain groups or outcomes. Institutions sometimes have a direct effect on policy outcomes; for example, governance systems with many veto points often have a bias favoring the status quo. More commonly, however, institutions interact with local political conditions or problem conditions to mediate the effect of these forces on public policy. Institutional design can affect how a government responds to population growth, financial strain, or heterogeneity of local preferences. Some institutions may be more responsive than others to a vocal minority's demands, or they may provide an easier path to elected office for members of underrepresented groups. A classic treatment of local government organization argued that "political institutions 'filter' the process of converting inputs into outputs" (Lineberry and Fowler 1967, 715). Subsequent scholarship has applied this framework to demonstrate how city and county institutions interact with political and problem context to shape local outcomes (Clingermayer and Feiock 2001; Lubell, Feiock, and Ramirez 2005; Mullin, Peele, and Cain 2004; Schneider and Teske 1995; Sharp 2002). This study uses a similar approach in comparing policies obtained through provision of services by a municipality and provision by an independent special district. I argue that the effect of institutional design is contingent on the nature and intensity of local demands.

Specialization and Responsiveness

The conditional theory of specialized governance begins with the assumption that special district officials have objectives that are similar to other public officials' objectives. They aspire to maintain their office through reelection or reappointment, and they also share with general-purpose legislators an interest in promoting policy-related goals. Their focus on a single dimension of local policy, however, affects how these motivations operate. Whereas city and county officials divide their attention among a diverse set of demands, specialized venues reduce the policy space to a single function or issue. Special district officials face lower opportunity costs for investing time to learn about technical complexities and political interests related to that issue. They also face limits on their ability to enact policy that crosses issue boundaries. In contrast, the constraints on cities and counties are informal: these governments have legal authority to act, but they may lack information due to demands on their time and attention. An important factor affecting the influence of these formal and informal constraints on politicians' behavior is the severity of the problem that governments are trying to address.

Objective conditions related to a public problem might influence local officials' behavior in two different ways. First, the severity of a problem should have differential impact on governments' capacity to search for and enact efficient policy solutions. Policy-oriented local officials should seek to make optimal use of public resources. Limits on time and information make doing so more difficult in a multidimensional issue environment, and the actual severity of public problems is likely to be a factor contributing to how general-purpose legislators allocate their attention. Special district officials do not face the same demands on their attention, and therefore they should be able to pursue efficient policies regardless of the status of problems within their issue domain.

Problem status also should have a differential impact on the electoral incentives facing specialized and general-purpose officials. Studies of policy responsiveness at the federal level have shown that representatives are more likely to heed constituent opinion on salient issues because those issues play a larger role in individuals' vote choices (Kingdon 1973; Page and Shapiro 1983; Wlezien 2004). Legislators also will attempt to anticipate constituents' preferences about potentially salient policies. A policy's potential salience is related to the severity of the problem and the likely distribution of the policy's costs and benefits (Arnold 1990). Issues are more likely to become part of the voting calculation when a problem becomes more severe or when policy decisions impose visible costs on constituents.

Thus, for city and county officials the electoral payoff for addressing a public problem varies with the seriousness of that problem.[22] General-purpose legislators have an incentive to respond to their constituents' perceived preferences on issues that are salient or potentially salient. Where objective conditions provide no reason for public concern, city and county officials perceive that the issue has low salience, leaving them free to follow their private preferences or respond to appeals by vocal minorities and interested groups. They anticipate that the median voter will pay more attention to policy actions in other issue domains. The severity of a problem should have less influence in a specialized policy venue that focuses on a single issue. Functional specialization reduces the voting space to a single dimension, so special district officials have reason to respond to constituent opinion even if the function they oversee is less salient than other local issues and if objective conditions are good. Voters have no other criteria on which to evaluate specialized politicians' performance, so responsiveness should be high regardless of issue context. Minority interests should have less ability to influence policy decisions, because special district officials have no other issues on which to earn majority constituent support.

This argument about the conditional relationship between specialization and responsiveness applies the concept of issue unbundling developed in papers by Timothy Besley and Stephen Coate (2002, 2003) on citizens' initiatives and elected regulators (see also Besley and Case 2003). Besley and Coate explore why majoritarian institutions might produce different policy outcomes when candidates always have an electoral interest in responding to majority opinion. In a general-purpose legislature, the bundling of multiple issues can yield nonmajoritarian outcomes on individual issues. Ballot initiatives and direct election of regulators change the incentives for policymakers and force greater policy responsiveness.

Special districts provide an alternative institutional mechanism for unbundling. In multidimensional elections for city and county officials, citizens have only a single vote to express preferences on a bundle of local issues. They select a candidate based on candidate positions on the most salient issues. For issue areas where objective conditions are perceived to be good, special-interest investments in campaign contributions and lobbying can distort policy outcomes away from the majority's preferences. Unbundling these issues from other dimensions of local policy should bring policy decisions closer to majority preferences because voters will evaluate special district candidates for their positions only on the individual issue. It also should reduce the logrolling that shifts policies away from the majority-preferred outcome on any one issue. For issues where problem conditions are poor, all local officials will have an incentive to respond to citizen preferences, so issue unbundling is less important. Figure 2.1 shows a graphic depiction of the argument.

Note that policy outcomes are predicted to be similar across institutional venues where problem status is more serious, but it is possible that general-purpose governments may be the less-biased venues under these conditions because of cities and counties' higher visibility and their more effective mechanisms for responsiveness. Many special districts are indeed hidden governments, operating outside the public view and without direct electoral accountability. Where problem conditions are severe

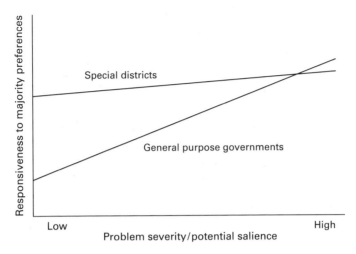

Figure 2.1
The effect of problem severity on responsiveness.

and general-purpose politicians have incentives for majoritarian responsiveness, their higher visibility and more extensive legal authority should promote more effective policymaking. Moreover, we should see variation in decision-making bias among special districts based on election rules that affect districts' structural capacity for responsiveness.

The conditional theory produces straightforward predictions about the effect of specialization on outcomes when a policy question is specific to a single local function. In the case of drinking water, questions about the level or structure of a pricing system or the adoption of new water treatment technology have few consequences that spill over into other functional areas.[23] For these single-issue questions, I expect problem severity to condition the effect of specialized governance by shaping the political incentives facing public officials in a general-purpose legislature.

The dynamics of specialization are more complicated for decision making on policy questions that involve multiple functions. Drinking water policy is inextricably linked with other local issues. The siting of roads and parks affects water quality and supply, and the design of a water system helps determine the efficacy of a community's fire protection.[24] Most important are the interrelationships between a community's approach to land use and the local water supply. Adequate water is a necessary precondition for new development; even if the planned homes and businesses will rely on private wells, developers and future homeowners need assurance that groundwater resources will support the proposed project. If the development will connect to a public water system, an investment in infrastructure and treatment facilities may be required. Not only does water supply affect options for land use, but the relationship also operates in reverse. Expanding a utility's service area can affect water quality and supply for existing customers and stimulate further development. The design of new development can lock in water-use patterns for decades to come because in many regions lawn care alone accounts for half or more of residential water use (U.S. EPA 2006).

General-purpose legislatures are equipped to address complex problems that involve multiple issues; the committee systems that most legislatures employ to divide their labor allow the development of issue expertise without sacrificing coordination and information exchange across issue areas.[25] Special districts are less flexible than traditional cities and counties with respect to their functional responsibilities, which creates challenges when addressing problems whose effects spill over functional boundaries. Water districts, despite their limited functional scope, frequently face policy decisions with impacts that reach into other issue domains—and into other local governments' jurisdictions.

I argue that special districts caught between formal limits on their authority and public demands for policy action will not restrict themselves to a narrow interpretation of their functional responsibilities. They will take action on policy questions that involve multiple dimensions. District officials have several incentives to stretch the functional boundaries of their authority. First, if the policy question has any relationship to the district's formal jurisdiction, constituents might expect district action—even if the policy itself lies outside the district's responsibility. Second, a further electoral incentive might exist if a district official hopes to use her position as a stepping stone to higher office. A politically ambitious official will want to build a policy record in multiple issue areas. Finally, district officials may choose to act based on their private preferences on the policy question, especially if those preferences coincide with their constituents' preferences.

Problem severity also plays a role in conditioning the impact of specialization on these complex policy questions. As with other types of policy decision, general-purpose governments will be more responsive to majority opinion where conditions related to the underlying problem are severe. Moreover, I expect problem severity to affect special districts by reducing districts' policy activity on questions that lie at the boundaries of their functional jurisdiction. Where problem conditions are normal, special district officials should feel free to follow their personal preferences or those of their constituents and implement policies whose impacts spill over into other issues. Problem severity raises the stakes on policymaking and expands participation by interest groups and the general public. Action by a special district entails more risk and may provoke a neighboring government or interested stakeholder. As a result, I predict that higher levels of problem severity make it more likely that districts will defer to the general-purpose governments' policy authority.

Specialization and Policy Coordination
The conditional theory proposes that problem context and institutional structure condition the effect of specialized governance not only on policy responsiveness, but also on cooperation among localities. Interlocal cooperation offers a wealth of benefits for local governments and residents. It can enhance efficiency by taking advantage of slack resources for the public good production and by providing compensation for negative externalities. A community that has reached the limit in its water treatment plant's capacity might contract with a neighbor for treatment services rather than build a new facility; a community that needs to protect its source water might pay a neighboring jurisdiction to prevent development in critical parts of a watershed. Cooperation also can help overcome service inequalities that emerge in a fragmented political system. Contracts and informal "handshake" agreements between governments allow small jurisdictions to pool their resources or obtain goods or services from larger local governments (G. Miller 1981; Stein 1990).

The scale of many special districts, their overlapping boundaries, and their limited functional scope combine to produce more opportunities for interlocal cooperation than in a system of general-purpose local governance. Special districts may have boundaries that were designed to address service gaps rather than to maximize efficiency in public good production. Some districts are so small that the investment costs for capital-intensive projects become prohibitive. The average staff size for a water district is just eight full-time employees.[26] Those districts with small staffs may have difficulty producing and delivering public goods and services independently. At the same time, cooperative agreements allow large special districts that are performing capital-intensive functions to reduce waste by sharing their excess capacity with nearby communities.

Interlocal cooperation offers even greater benefit when a policy's impacts reach across issue boundaries, so that a district's policy decisions have effects that spill over into other governments' jurisdictions. Take, for example, a water utility that is planning for the replacement of aging pipes. The project may be necessary for long-term reliability in water service, but it will have negative spillover effects in the short term by disrupting traffic, creating noise and disturbance for neighbors, and imposing losses on commercial businesses. If the utility is part of a general-purpose local government, the city or county might lessen these costs by allocating more traffic control and street repair resources to the affected neighborhoods and by coordinating pipe replacement with other infrastructure renewal projects. A consolidated jurisdiction would act on its own to minimize the negative spillovers. A water district would have less ability to redirect resources in this manner, making contracts a more attractive option for managing spillover effects. It would be less costly to contract for traffic control and street repair than to keep personnel on staff to perform these duties. Thus, because of territorial overlap and narrow functional scope, special districts may enjoy particular benefits from pursuing alternative service arrangements through cooperative partnerships.

Along with greater benefits, however, come higher costs. Whereas a city council would designate the contributions of various municipal departments in the case described, specialized governance requires agreement among autonomous entities in order to lessen the impacts of the pipe-replacement project. Bargaining between independent actors is costly under any circumstances due to uncertainty and the potential for opportunistic behavior (Williamson 1975). These costs are even higher for political institutions, because the weak linkages between representatives and their constituents heighten uncertainty in the enforcement of agreements (North 1990). Politicians may negotiate agreements without full information about citizens' preferences. When these agreements are among departments within a city or county structure, hierarchical relationships provide little opportunity for departments to renege. Negotiation between governments shifts debate into the public sphere, introducing new actors who might try to influence the elements of a partnership agreement or to exercise a veto. In effect, negotiation expands the scope of the conflict and increases uncertainty about its outcome (Schattschneider 1960). I argue that specialization has the potential to make policy coordination more visible to constituents but also more costly.

Interlocal cooperation is a form of institutional collective action that may allow achievement of collective benefits that cannot be attained by governments acting independently.[27] Under the right conditions, agreements can make all parties better off by using existing resources more efficiently and making it easier for local governments to satisfy their constituents' preferences. But information constraints, bargaining costs, and enforcement problems can interfere with the development of cooperative relationships. Institutions are most likely to overcome the hurdles to collective action where policy problems are severe and resources are available to help offset the transaction costs associated with negotiating and reaching agreement (Lubell, Schneider, Scholz, et al. 2002). Public choice and metropolitan reform theories offer different assessments of these costs. Public choice assumes that transaction costs are low and treats interlocal contracting as a critical tool for achieving efficiencies in a competitive public economy. Metropolitan reformers perceive the costs as prohibitive and therefore predict little cooperative activity to emerge in a fragmented political system.

Here again, I argue that attention to policy context is critical for understanding the impact of institutional design. A system of specialized governance offers more opportunities for interlocal cooperation, but the costs of engaging in cooperation are high. Factors that influence the benefits or costs of cooperation will affect the likelihood that cooperative behavior emerges. One such factor is problem severity, which increases the benefits of overcoming the hurdles to collective action. Where problems are severe, politicians will come under pressure from constituents to find a policy solution, which will make them more likely to see the advantages of developing partnerships with neighbors. The political and policy benefits that accrue from addressing a serious problem help to outweigh the costs of conflict expansion.

Another factor is special districts' institutional structure, which can affect the incentives for district officials to pursue cooperative relationships. Structures that insulate officials from public accountability, such as appointment of district officials, reduce the potential benefits from solving an important policy problem, but at the same time they also might lower the visibility and consequently the costs of negotiation.[28] The flexibility of special district boundaries may influence cooperative behavior by reducing the benefits of interlocal agreements. Special districts that can easily adjust their territorial boundaries to address spillovers or respond to changing patterns of demand should be less likely to engage in partnerships in order to achieve these efficiencies. Finally, special district boundaries that crosscut those of neighboring cities and counties create distinct constituencies that may have diverging preferences on complex policy issues. These conditions will make it more difficult for policymakers to reach agreement and may introduce more actors into the policy process. In contrast, contiguity between special district and general-purpose boundaries should reduce the scope of conflict and provide more favorable conditions for cooperation.

Specialized Governance of Drinking Water
The debate over the costs and benefits of fragmentation in metropolitan governance has carried on for half a century.[29] Less visible but nearly as long lasting is a similar debate over fragmentation in the governance of water resources (Blomquist 1992; Godwin, Ingram, and Mann 1985). Both debates have centered on the relative benefits of competition and centralized hierarchy, dedicating less attention to the consequences of government specialization. This book takes on the latter task.

In the chapters that follow, I examine who benefits from a system of specialized governance. A responsive government should benefit its constituents by delivering policies that a majority of constituents prefer. Not everyone wins in a majoritarian system; citizens with outlying preferences may be left unsatisfied. Yet the alternative is a biased system that consistently favors the interests of a minority over those of the majority. A government that collaborates with its neighbors benefits constituents by delivering services more efficiently and helping develop effective policies to promote regional integration and equality.

It is the public who benefits from responsiveness and interlocal cooperation. A biased system favors a minority, usually wealthy, highly educated homeowners who have higher rates of political participation (Fischel 2001; Gilens 2005; Oliver and Ha 2007). In the context of local politics, another minority that stands to gain is made up of the developers and landowners who profit from residential growth. According to the influential growth-machine hypothesis (Logan and Molotch 1987; Molotch 1976), the default position for most local governments is to join economic interests in pursuit of growth. Individuals and businesses who stand to profit from development decisions invest time and money to influence the local officials who make those decisions. They also will take advantage of long-standing and persuasive symbolic arguments about the positive benefits of growth for a community. Residents who seek to organize against growth must overcome collective action problems (Olson 1965; Schneider and Teske 1995) and systemic conditions that predispose local officials to ally themselves with the growth machine (Stone 1980). The antigrowth activists are often resource-rich homeowners, however, who themselves may enjoy access to the community's decision makers.

This study evaluates whether specialized governance provides an advantage to either of these groups. The focus is on identifying and measuring the effects of specialization, but in doing so I propose a new way of looking at policymaking in traditional city and county venues. If the conditional theory holds, then developers and resource-rich interest groups often do have disproportionate influence over municipal decisions about development and urban service delivery. Politicians in general-purpose venues concentrate on satisfying constituents on the most salient issues, perhaps schools or public safety, and they respond to the growth machine on issues that residents ignore. But the conditional theory also helps explain situations where a growth machine fails to dominate city or county policymaking. When policy challenges become more serious and citizens turn their attention to an issue, city and county politicians are not willing to pay the potential electoral price for developer domination.

The empirical tests examine contemporary policy challenges in the management of drinking water. In some respects, we should expect to see less variation in preferences about water policy than about many other local issues. It seems safe to assume that everyone wants access to a clean, reliable source of drinking water at a low price. Although this goal may be near universal, specific policy decisions present trade-offs involving willingness to pay for public goods, the distribution of service costs, the choice of spending priorities, and local governments' responsibility to accommodate and provide for residential growth. Indeed, Paul Peterson (1981) declares that allocational functions such as the provision of drinking water—local services from which all residents benefit—give rise to the most contentious local politics, because one resident's gain is likely to impose a cost on another resident. In a biased system, some groups consistently gain despite being in the minority. The empirical analyses that follow look for evidence of bias across institutional structures and problem conditions. If specialized governance provides a systematic advantage to certain interests in the development of water policy, there is reason for concern that the bias extends to other policy areas as well.

Apart from detecting patterns of bias and responsiveness, investigation of drinking water policy reveals whether the nation's public water utilities are prepared to address the challenges associated with distributing a scarce resource among a growing population. A biased system can exacerbate socioeconomic inequality and diminish trust and confidence in governmental authority. But our requirements for local governance extend beyond responsive policymaking, especially for a function as essential and as vulnerable as the provision of drinking water. Governments also must be competent in developing and implementing policies that will achieve efficiencies and extend scarce resources. They must be receptive to policy innovation and be willing to cooperate with neighbors. The new local water politics places high demands on local governance, and we need to understand whether governing institutions are ready to confront the challenge.

The Question of Institutional Choice

The empirical analyses in the following chapters take institutional design as a previously determined choice and examine its consequences for public policy. Before I turn to the analyses, some words of defense for treating institutional design in this way are in order. Of course, the choice of specialized governance is not strictly exogenous to the local political and problem conditions I address. Political actors create special districts in order to solve problems and help satisfy local demands. If these actors are making a simultaneous decision about institutional design and public policy, the relationships I uncover between special district governance and policy decisions may be spurious—factors related to the local context may account for both outcomes.

A full accounting of institutional choice is outside the scope of this study. However, there are two dominant reasons why the determinants of institutional design cannot explain the policy choices documented in my analyses. First is an issue of timing. On the whole, water districts are old relative to the policies under consideration in the new local politics of water. The policies I examine are strategies that have emerged in the past twenty-five years for pricing and distributing water in the face of resource constraints. Most water districts were formed earlier, during the era of supply-side solutions. Starting with a national sample of water utilities surveyed by the American Water Works Association (AWWA) in 1999, I used information from the U.S. Census Bureau's *Census of Governments* and from state and water district Web sites to identify dates of district formation. Within the sample of 107 water districts for which information was available, the median incorporation date was 1953. More than three-quarters of the districts formed before 1970, and just thirteen districts formed after 1980.[30] In most communities, problem conditions related to water persisted between institutional design and the recent policy adoptions that I examine. The organization of local political interests almost certainly changed, however, especially considering the dramatic shift that occurred in policy responses to water scarcity during that time. Previous research on special districts suggests that developers often dominated district formation, especially during the 1950s and 1960s (Burns 1994); developers are the leading detractors of the policies discussed here. If local actors are jointly choosing specialized governance and water-policy outcomes, we at least should expect more recently formed water districts to be more likely to adopt these policies. In the analytic chapters, I provide evidence demonstrating that this pattern does not hold.

Second, only a weak relationship exists between local problem conditions and institutional design. Although specialized water governance is more common where water problems are more severe, water districts in fact are scattered throughout the country. Table 2.2 shows data from the

Table 2.2
Water Districts by Region, 2002

	Number of water districts	Number of water districts per 10,000 population
Northeast (CT, ME, MA, NH, NJ, NY, PA, RI, VT)	4,680	0.87
Midwest (IA, IL, IN, KS, MI, MN, MO, ND, NE, OH, SD, WI)	12,229	1.88
South (AL, AR, DE, FL, GA, KY, LA, MD, MS, NC, OK, SC, TN, TX, VA, WV)	8,443	0.82
West (AK, AZ, CA, CO, HI, ID, MT, NM, NV, OR, UT, WA, WY)	9,700	1.48

Source: U.S. Census Bureau 2002.

2002 *Census of Governments* (U.S. Census Bureau 2002b) on the number of water districts by region. The West has a large number of water districts relative to the nation as a whole, but the middle and upper Midwest is even more district reliant, even though mean temperatures are low, precipitation is typically more than adequate, and water supplies are plentiful. The South has the smallest number of water districts per capita, even though it includes Texas and Oklahoma, two of the most district-reliant states.

At the local level, climate variables that are strong determinants of water demand have little power in explaining institutional design. To measure the effect of water problem severity on the choice of specialized governance, I plotted the utilities in the AWWA sample on a map of the nation and identified local climate characteristics for each utility's location.[31] Indeed, both the average maximum daily temperature in a location and the average level of precipitation have a statistically significant relationship with institutional organization in the expected direction. Water districts are more common where the weather is hot and dry. However, knowledge of local problem conditions helps little in predicting the form of a community's water governance. A model of institutional choice that includes these climate variables produces less than a one percent reduction in error in predicting governance type over a null model predicting that a community will have a city- or county-operated water utility simply because that is the most common governing arrangement. Moreover, existing research demonstrates that district formation depends in large part on the influence of state enabling legislation, which may not be related to local problem conditions.

In sum, although water scarcity might promote the establishment of specialized institutions for managing public water supply, water districts are actually common in places where supply is plentiful, and the factors influencing the choice of specialized governance at the time of community formation are unlikely to account for contemporary choices over water policy. By treating special district governance as exogenous, I assume that no unmeasured factor influencing the policy outcomes I examine is correlated with the assignment of water districts. Instead, specialization and the crosscutting political boundaries that go along with special district governance create their own political incentives, changing local interests' political strategies and helping shape the policy decisions made by public officials.

3
Private Costs and Public Benefits in Local Public Services

The most visible policy decisions a local water system makes are decisions about prices. Most citizens interact with their water provider only when they pay their bill. However, even careful scrutiny of a water bill will miss some of the critical aspects of a pricing policy. Pricing is a means for water providers not only to recover the costs of system operations, but also to create benefits for certain customer classes, offer incentives for specific types of water use, or provide a public benefit. The power of pricing as a policy instrument, combined with its public visibility, makes it a good place to start in exploring the policy effects of specialization in water governance.

This chapter examines a specific type of water rate policy that offers a number of public benefits, including water conservation and income redistribution, while imposing concentrated costs on a community's wealthiest residents. The policy has gained popularity among environmental advocates, water-resource managers, and economists as a strategy for encouraging efficient water use. The chapter presents a series of analyses measuring the effect of institutional organization on a utility's likelihood of adopting this policy. The analyses assess the influence of specialized governance on the choice of water-management strategies and more generally provide an opportunity to test whether special districts are more biased toward private interests than are traditional cities and counties. The results demonstrate how governing structure shapes incentives for local officials to provide public goods in the face of attempts by private actors to veto those policies. The effect of specialization is not universal, however; the provision of public goods from general-purpose venues is conditional on the severity of water scarcity, but special districts' policy choices appear to be independent of problem context. Institutional factors related to the selection of district officials further shape the incentives for a water system to provide public goods.

The Nature of Drinking Water as a Good

In the typology of goods used by economists, water does not fall neatly into any single category. One distinction economists make is whether a good is rivalrous, so that consumption by one user precludes simultaneous consumption by someone else. When water is used for navigation, wildlife habitat, or even most recreational purposes, its use is nonrivalrous; many users can share its benefits. In that natural context, water also is nonexcludable because one user cannot prevent another from enjoying a sailing trip or the rewards of ecological diversity. These two features—nonrivalry and nonexcludability—are the characteristics of a public good. Because public goods are shared, and the individual incentives to provide them are weak, public goods typically are undervalued and undersupplied.

Water has different attributes in other contexts. If the intended use is for irrigation, industry, or drinking water, then consumption becomes rivalrous. Water in this context is a common pool resource. It becomes vulnerable to overuse, because without imposition of a governing institution, it is difficult to put limits on extraction of water that is stored in a river or groundwater aquifer. Once water has been treated and distributed for use as drinking water, it becomes a private good. Exclusion is feasible, and user charges can be applied. In general, markets do an excellent job at providing private goods. But local water service retains features that are related to public goods: water systems must protect the quality of their water sources and maintain adequate pressure for firefighting. Absent regulation, consumers and utilities tend to undervalue these benefits of a public water system.

Complicating things further is the fact that water is an essential good. Even if exclusion is feasible, it is not desirable, and there are no substitutes for clean, safe drinking water. Finally, the cost structure of a water system differs from that of most local services in that water provision is highly capital intensive. A large proportion of the expense in providing drinking water comes in fixed costs, and the marginal cost of providing an extra gallon is low, especially in the short term. Water systems frequently are built with excess capacity, so there may be little true rivalry in consumption except in times of drought.

The history of drinking water provision in the United States reflects these complexities in the nature of the good. Over the course of the nineteenth century, industrialization and urbanization prompted demand for the formation of community waterworks to replace individual water collection and extraction. Private firms responded to this demand and dominated the early development of water systems, just as they did for other local utilities. It was not long, however, before municipalities took on a bigger role, and by 1896 the majority of community waterworks were publicly owned and operated (Crocker and Masten 2002). The trend toward public water provision continued into the twentieth century, eventually leaving private water firms with control over only a small minority of the nation's large water systems. Yet ownership of other utilities, such as for gas and electric power, remained in private hands. An important reason for the transformation in management of drinking water relates to the public benefits of the good; some private water companies' failure to provide public-health protection and adequate flow for fighting fires contributed to public demands for municipal takeover. Cities seem to have understood the positive externalities that could follow from maintenance of a high-volume, clean water supply, and so they increasingly took on the task of building water systems, commonly subsidizing municipal water operations out of tax revenues.

Cross-subsidies were necessary because it was difficult to raise enough revenue through user fees to cover the costs of water-system operation.[1] Fees based on actual consumption were rare, because most water utilities did not require metering. Although metering for water was no more costly than for gas and electricity, the high fixed to marginal cost ratio for water provision made metering less economically viable than for these other utility services. Flat fees for access to the system also were impractical because effective use of access fees would require excludability. Excluding households from water service was technologically feasible, but it imposed substantial public-health costs. Municipal utilities therefore had an incentive to keep access fees low so that users would not opt out of water and sewer service altogether.

Thus, since the establishment of community water systems in the United States, their role in providing public benefits in addition to a private good has had an important impact on their governance and revenue policies. The spread of water metering provided further opportunity for utilities to use pricing to promote a public good. In charging directly for water consumption, utilities can design rate systems that provide incentives for water conservation, income redistribution, or economic growth. They can send signals to coordinate collective behavior regarding levels and timing of water use. Pricing is a powerful policy tool, and using prices to pursue public goods may create tension with some residents' private demands. Water rate structures differentially affect customers according to water uses and patterns of residential development, and they reflect trade-offs among equity, revenue stability, efficiency, and administrative practicality. These aspects of local water provision can give rise to disputes among residents who have different consumption demands or preferences about public goods. As William Berry notes regarding electricity rates, "the setting of rate structures is inherently redistributive; its study, therefore, can extend our understanding of policy making which reflects a conflict for benefits between 'haves' and 'have-nots'" (1979, 263). Policy choices regarding water prices have consequences for who pays for and who benefits from local services. Therefore, they provide a useful context in which to test for bias in policymaking.

The Public Benefits of Increasing Block Rates

Water rate structures vary in how they distribute the costs of water service, both across and within residential, commercial, and industrial customer classes.[2] There are four primary water rate structures: (1) a flat-fee system that charges the same sum to all customers, regardless of the amount of water consumed; (2) a uniform rate with the same marginal price for water at all levels of use; (3) a declining block rate in which the per unit rate for water declines with increased consumption; and (4) an increasing block rate in which water costs more per unit as consumption rises.[3]

The revenue choices of public water systems have adapted over time to technological improvement and changing social and industry values. The introduction of water metering allowed utilities to adopt pricing schemes that account for the quantity of water consumed. Although metering is not universal and many water systems do not require existing housing stock to be retrofitted with meters, it is now rare for a utility to charge a single flat rate for all customers. Some communities actively resist the installation of meters, even in regions where water scarcity is a severe problem. In 2003, local officials in Fresno, California, supported state legislation to mandate residential metering in their community after failing to convince voters to overturn the local ban on metering, even as the federal government threatened to cancel the city's water supply contract. Sacramento successfully resisted state efforts to overturn its city charter's ban on meters until 2004. In general, however, water systems have taken advantage of the opportunity to impose charges according to water usage.

As utilities began to meter residential water consumption, many of them adopted uniform rate systems that were simple to design and administer. In the 1960s, a number of systems began to implement declining block rates based on embedded-cost principles. In order to assure reliability of water supply and meet variable demand, utilities must build a substantial amount of excess capacity into storage and treatment facilities. Most of the time, this excess capacity is not required to meet customer demand. Because large-volume water customers have fairly constant demand, the unit cost of providing capacity to meet their water needs is lower than for residential and small commercial customers, whose demand varies across seasons and times of day. Utilities sought to adopt a single rate structure for all customer classes that would recover current operating and capital costs, and declining block rates most closely matched the short-term costs of service by allocating the costs of excess capacity to customers with more variable demand.

Uniform and declining block systems were until recently the dominant rate structures in the water industry. In the late 1970s, the principles behind declining block prices came under fire across utility sectors. The trend began in electricity, where inflation and the energy crisis produced dramatic price increases and fuel shortages that prompted efforts to promote energy conservation. The 1978 Public Utilities Regulatory Policies Act required private electric utilities to justify continued use of a declining block structure over other revenue policies, and state utility regulators began to consider economic efficiency in approving utility pricing plans for electricity.

These developments in the electricity sector raised questions about declining block rates among utility industry professionals, and many water systems soon began to reconsider their own use of declining block prices. Growing concerns about water supply and complaints by consumer advocates about the fairness of declining block rates to residential customers provided further reason to find an alternative rate structure. In the 1980s and 1990s, water shortages, environmental regulation, and deteriorating water infrastructure produced marked increases in water rates throughout the nation, heightening attention to fairness in the distribution of utility costs (AWWA 2000, 270).[4] Water prices historically have been low in the United States, and local politicians and residents were willing to defer to water experts in designing rate structures as long as water remained affordable.[5] Rising prices increased the salience of water rate design as a local political issue, creating demand for more transparent and participatory rate-setting processes. As a manual providing guidance on developing conservation rate structures explains, "Many water agencies have realized that ratemaking no longer is a mere technical exercise. Real increases in the price of water and the likelihood of water shortages have thrust the process of designing rates into the harsh light of public scrutiny" (Chestnutt, Beecher, Mann, et al. 1997, 3-1). A changing regulatory climate and public dissatisfaction with rates that penalize residential users created pressure on water utilities to cease their reliance on declining block prices. Some switched to uniform rates, and others adopted increasing block structures in which water costs more per unit as consumption rises.

Figure 3.1 shows the shift in residential rate structures from 1982 to 1997 among utilities located in major metropolitan areas.[6] Over that time, the percentage of sampled water systems whose prices declined with consumption decreased by twenty-six points, and use of increasing block rates spread from just 4 percent to 31 percent of all water systems. Like other rate structures, increasing block pricing can be designed to

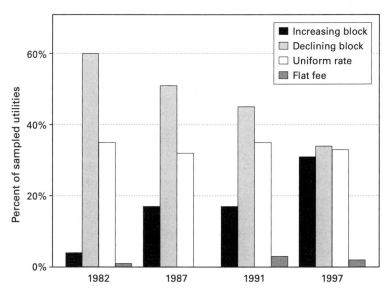

Figure 3.1
Residential water rate structures, 1982–1997. *Source:* Organization for Economic Cooperation and Development 1999.

generate any level of income. One drawback to an increasing block plan is that revenue will be more variable; when the marginal units of water consumed are those whose prices are highest, reduced consumption due to weather or conservation investments creates larger fluctuations in overall revenue. However, careful consideration about the number of blocks in a progressive rate structure, the levels of block switch points, and the price differentials between blocks can increase revenue stability. Moreover, increasing block prices offer a number of public benefits that are absent in alternative pricing schemes: economic efficiency, water conservation, and redistribution.

Economists have long promoted the adoption of increasing block rates as a means to enhance efficiency, arguing that consumers should receive price signals that indicate the increasing marginal cost of providing water service. In markets such as water where costs are rising over the long run, rate structures based on historical average costs contribute to excessive demand and inefficient investment in new supply. True marginal-cost pricing is neither technically nor politically feasible in the public drinking water sector, but increasing block rates represent an efficiency gain over uniform or decreasing block prices by signaling the increasing incremental costs of water provision.[7]

By providing a price incentive to discourage wasteful consumption, increasing block rates also can be used to promote water conservation. Research has shown that water demand can be responsive to price (Hewitt and Hanemann 1995), especially in the presence of a block rate structure (Dalhuisen, Florax, de Groot, et al. 2003; Olmstead, Hanemann, and Stavins 2005).[8] Although economists continue to debate the precise size and form of block pricing's impact on water use, policymakers at the state and federal levels are convinced of its efficacy as a conservation strategy.[9] Reducing overall water consumption offers a number of public good benefits, such as preventing subsidence and saltwater intrusion, improving water quality, and maximizing in-stream flow for recreation and habitat protection. Conservation can promote equity and productivity in resource allocation by redistributing scarce water supplies from unnecessary to essential uses. Even in the absence of scarce supply, reducing water demand may allow utilities to delay or avoid the capital costs of new storage and treatment facilities. In sum, the benefits of conservation are widely shared within and across communities, and block pricing is less severe than rationing and other alternative policy instruments for achieving water-use reductions.

Finally, progressive rates can promote vertical equity and income redistribution.[10] Within a given set of climatic conditions, income is the strongest determinant of household water demand.[11] As income rises, households use more water to operate appliances, fill swimming pools, and maintain green landscaping on larger lawns. The highest price tiers of an increasing block structure affect only those households that consume the most water. Every household's basic needs are subsidized under an increasing block plan, and low-income households whose demand is restricted largely to indoor use pay the lowest price for water.[12] With a declining block system, the highest rate applies to all households for their essential uses, and high-income customers pay less for the increased consumption attributable primarily to outdoor uses.[13]

Utilities can use lifeline plans and other methods to relieve the burden of water rates for low-income customers, but the increasing block plan builds equity considerations directly into the rate structure by minimizing the difference across income groups in the proportion of household expenditures dedicated to water.[14] Market-based pricing tools are often more politically feasible than direct subsidies for promoting affordability, and the historical underpricing of urban drinking water suggests that local politicians may have an incentive to manipulate water prices in order to pursue income redistribution (Timmins 2002).[15] Increasing block rates favor essential over nonessential consumption for residential customers and therefore subsidize customers with the least disposable income. By shifting the burden of water costs to large users, however, progressive prices are likely to attract opposition from the most politically active members of the community.

Considering the distributional implications of increasing block rate structures, it is not surprising that in many communities their adoption has met with controversy.[16] The Los Angeles Department of Water and Power's 1993 decision to adopt progressive rates prompted citizen protests in the San Fernando Valley after residents saw marked increases in their water bills. Valley residents voted a city council member out of office based on the rate change, and some residents threatened lawsuits and even made death threats (Hall 2000). Citizens in Fort Collins, Colorado, clashed for several years over a tiered rate structure that the city had adopted in response to drought conditions in 2002. Residents with large lots proclaimed the rate structure unfair and called for a return to uniform pricing once the drought ended. In 2005, a new council majority nearly abandoned tiered rates, but it kept the system in place to avoid jeopardizing the city's federal permit application for a reservoir expansion project (Benson 2003, 2004, 2005; "Halligan Project" 2006). In both of these cases, the utility ultimately kept the progressive rate system but adjusted some of the rate levels. However, the cases demonstrate the political risks associated with shifting to an increasing block rate system. Rate manuals now dedicate considerable attention to advising water systems on how to build public support for a rate proposal (AWWA 2000; Chestnutt, Beecher, Mann, et al. 1997).

The emphasis on public acceptance of a rate schedule is new for the water industry, which for a long time disregarded the idea that water pricing could be used for any purpose apart from cost recovery. The industry's changing approach to the pricing of drinking water is evident in the guidance provided by the leading industry manual on rate setting, which historically advised strict adherence to embedded-cost principles that reward the stable demand characteristics of large customers (AWWA 1991). The manual recommended designing rates to match as nearly as possible the current average cost of providing service to each customer, without regard to any future cost of system expansion. The result was a rate system that charged the highest per unit cost to customers with the most variable water demand—residential and small commercial users. It discouraged using any other criteria in developing rates, specifically rejecting "substantial departure from cost-of-service-based rates to achieve social objectives" (AWWA 1991, ix). The manual viewed the rate-making process as a technical exercise and treated political factors as an intrusion into what should be an expert-dominated process. As recently as 1991, it criticized the spread of increasing block prices (also known as *inverted rates*) because they "do not recognize the generally better load factors of the large-use customers.... A very practical objection to inverted rates is that higher use per customer does not necessarily indicate a higher cost per unit of use" (AWWA 1991, 50). The manual recommended that before adoption, increasing block rates be carefully evaluated for their relationship to service costs.

By 2000, water industry leaders had changed their outlook. The next edition of the manual adopted a more flexible approach to rate setting, recognizing that a utility might have different and multiple objectives and that rate structures can be used to meet those objectives:

A utility is presented with a major challenge when it sets out to select a rate structure that is responsive to the philosophy and objectives of both the utility and its community. It is important to the utility and its customers to select the appropriate rate structure because the majority of the utility's revenues are collected through water rates and because pricing policies may support a community's social, economic, political, and environmental concerns.... When diverse and competing objectives are well understood and evaluated, a utility has the opportunity to design a rate structure that does more than simply recover its costs. A properly selected rate structure should support and optimize a blend of various utility objectives and should work as a public information tool in communicating these objectives to customers. (AWWA 2000, 79)

The manual carefully considers the variety of rate structures available and judges each on multiple criteria, including revenue stability, ease of implementation, and equity within and across classes. It continues to defend declining block rates for reflecting the distribution of a water system's costs, but it acknowledges the trend away from declining block systems based on perceptions of inequity. It also offers a more positive assessment of increasing block rates than had the previous edition, noting that "properly designed increasing block rates recover class-specific, cost of service while sending a more conservation-oriented price signal to that class" (99). The water industry had come to recognize the power of pricing tools to pursue a variety of public benefits.

Adoption of Increasing Block Rate Structures

Progressive rates are a prime example of a policy that imposes private costs on a subset of citizens in order to pursue public benefits that will be shared widely.[17] By providing incentives for more efficient use of water resources, increasing block rates promote economic efficiency and water conservation. They are an important strategy for water managers trying to reduce waste and to increase the productivity of existing supplies. They also offer social benefits in that they protect affordability for essential water uses by imposing costs on excess usage, typically by the wealthiest members of a community. Local governments considering adoption of increasing block pricing likely will face opposition from those who bear the heaviest costs under the policy. Water rate policies thus offer a useful context in which to compare choices made by different governing structures when a trade-off exists between public goods and private interests.[18]

Hypotheses

In addition to their public good benefits, progressive rates—like progressive taxation—offer a direct pocketbook benefit to the median resident in most communities.[19] The distribution of household water demand has a strong right skew: the majority of households consume a quantity of water below the household average, and frequencies decline at higher consumption levels. Although the effect of a block-pricing system on an individual resident will be influenced by the number of blocks and the levels of rates themselves, any progressive rate structure will impose the largest per unit cost on high-income residents who consume water in the right tail of the distribution. Because the median household tends to fall into a category of low consumption, that household's water use will be subsidized by those with higher demand. One rate manual used single-family residential consumption data from several California water agencies to estimate the impact of a typical revenue-neutral change in rate structure from uniform to increasing block rates (Chestnutt, Beecher, Mann, et al. 1997).[20] In one example, 80 percent of customer bills declined. Approximately 15 percent of bills increased, with a maximum 30 percent escalation. Another example set the first block switch point so that 70 percent of households consumed entirely within the first block and therefore enjoyed a rate cut. Given the typical distribution of residential water use, the median household will pay less under an increasing block rate system than under a uniform or declining block rate system designed to generate the same revenue.

Because of the strong link between water use and income, the highest rates fall on the jurisdiction's wealthiest residents. People with large houses and large lots consume the most water, and under a progressive pricing scheme they subsidize the water use of a larger number of consumers. Indeed, opposition to increasing block rate plans typically comes from high-income residents and developers who profit from building expensive housing units, because the concentrated costs of a progressive pricing structure would fall on these groups. A manual on conservation rate designs highlights the importance of a "vocal minority" that might block implementation of a rate change:

> The vocal minority can be particularly effective as a blocking coalition because its size and membership allow it to make its needs and desires more keenly felt by decision makers than can the "silent" majority. Its relative small size lowers its organizational costs. More importantly, the risk of losing income discourages free-ridership making it very difficult for the silent majority to organize itself as a countervailing force to the vocal minority. Thus, the vocal minority may effectively prevent a change to a new rate structure even if the change would make the utility's customers as a whole better off. Conservation-oriented rates designed to discourage some type of discretionary use may be particularly susceptible to

actions by a vocal minority. Those significantly vested in the use in question are likely to resist the change. (Mitchell and Hanemann 1994, 73)

The model presented in the next section tests if the influence of this "vocal minority" varies across institutional settings.

The existing theories about special district responsiveness described in chapter 2 produce competing hypotheses about the effect of specialized governance on the decision to adopt increasing block pricing. If the conventional wisdom is correct that special districts are biased to favor developers and wealthy private interests, then special districts should be less likely than city and county utilities to adopt progressive pricing. Critics of specialized governance contend that special districts' low political profile favors actors who have substantial resources to invest in monitoring decisions and lobbying district officials. Those who would bear private costs under an increasing block pricing system have a more acute interest in the policy and more resources to invest, so they should have greater success in vetoing a rate-change proposal in circumstances where policymaking is less visible to the larger public.

Public choice theory would predict the opposite. According to this view, special districts are embedded in a competitive local public economy that rewards economic efficiency and policy responsiveness. Cities and counties that have a territorial monopoly for all or most public services do not face the same market and political incentives. Special districts should be more likely to take up a policy innovation that promises more efficient and productive use of water resources, especially when the innovation offers public good and pocketbook benefits to the median resident. Special districts allow citizens to separate their preferences on water rates from other local issues and to express their demands directly to water officials, producing greater responsiveness to median opinion. Moreover, the specialization and multiplication of policy venues provide a greater number of access points than in consolidated systems and reduce the costs of participation for those with limited political resources. Developers and high-income residents who have the most to lose from a shift to increasing block prices will have more influence in a traditional local government with fewer access points. In a traditional venue, residents who would receive a modest benefit from progressive pricing will spend their political resources to influence decisions on land use, public safety, or some other salient issue. In a specialized water district, however, water prices are the most salient policy question for nearly all residents. High-income water consumers will not enjoy the same procedural advantage as in a city or county venue, and the median resident's preference should be more likely to prevail.

I test these predictions against the conditional theory of specialization, which proposes that the effects of specialization will vary based on the objective conditions of water supply. The politics of water take on a different character across U.S. communities depending on water availability. Throughout much of the Northeast and Midwest, water provision is a low-profile function for local governments. Apart from occasional emergencies due to pollution or aging infrastructure, citizens in most communities take for granted the water that flows from their taps. It is in these communities that governing structure should matter most in shaping water policy. Water issues will be low on the agenda of city and county officials, allowing attentive groups to bias rate policy away from the majority's preferences. In the hotter, drier climates, where securing a reliable water supply has long been a central concern for governments at all levels, water consistently has a place on the public agenda. Public officials in all institutional settings recognize that citizens and interest groups may monitor decision making on water issues and incorporate water-policy decisions into their vote calculations. Here we should see little difference between policies delivered by special districts and those adopted by general-purpose governments.

Model

The model uses data from a 1999 survey of water utilities to estimate the influence of governing structure on the adoption of increasing block rates for residential users.[21] The close correlation between household income and water use attaches clear distributional consequences to a residential pricing strategy. The same is not true for commercial and industrial users, whose water use varies across industries. Moreover, among nonresidential users the distinction between public and private benefits is more complex, because economic development might compete with conservation and redistribution as a public good. In some industries, water makes up a substantial proportion of operating costs, and water availability and price can influence a company's decision to locate in an area. Rates that attract and keep large industrial users in the water system can provide jobs, tax revenues, and other public benefits. Local officials might choose to set water rates in order to pursue these benefits. By penalizing large users, increasing block rates may support provision of some public goods but interfere with others. In short, it is impossible to assume preference distributions or the imposition of costs around progressive rates for nonresidential customer classes, and thus this analysis does not treat these classes.

The dependent variable measures whether a utility has adopted an increasing block plan as part of its package of residential rate structures. Focusing only on increasing block rates is a decision motivated by theoretical and measurement considerations.[22] Theories about the relationship between specialization and responsiveness produce clear hypotheses about the adoption of progressive rates; the implications are less clear for decisions about other alternative rate structures. Moreover, use of increasing block rates reflects a relatively recent policy decision. The first water utilities in the United States to adopt these rates did so beginning in the late 1970s, and the innovation did not spread until the mid-1980s. Where the use of declining block rates rather than uniform rates might reflect a policy decision made many decades earlier, the use of increasing block prices indicates a contemporary policy choice to abandon an existing revenue policy. Restricting the analysis thus allows more direct measurement of the effects of institutional variables and minimizes the problem of using cross-sectional data.

The key explanatory variable measures whether the utility is managed by a special district or by a general-purpose local government.[23] In order to take account of other factors that might influence the choice of rate structures, I estimated a probit model that controls for system and financial characteristics of the sampled utilities. Included in the model are variables indicating a utility's dominant source of water supply, which affects the utility's cost structure. It is usually more expensive to obtain water from surface sources than from groundwater sources, and the price of purchased water depends on its original source and the availability of alternative supplies. Increasing block structures create more revenue variability, so utilities with greater water supply costs might be less likely to adopt these rate plans. A utility's operating ratio is an indicator of its financial health. It is not possible to measure characteristics of a utility's customer base such as its income distribution and partisan composition, because few states make available geographic data on special district boundaries that would allow calculation of demographic and political characteristics. A variable indicating whether the city or water district is located in an urban area helps account for differences in the distributions of water demand that might have an impact on rate-structure preferences.[24] The model also includes dummy variables that indicate utilities' regional location in order to account for aspects of political culture and historic water allocations that might affect the likelihood that a utility will adopt progressive rates. Utilities should dedicate more attention to retail rate structures when retail rather than wholesale water sales make up a greater proportion of their enterprise, so the model also controls for retail customer population as a percentage of the total population served.[25]

Data characterizing the local climate capture objective conditions related to water in a community. Precipitation and temperature are important predictors of both supply and use of water.[26] In hot, arid climates, water resources are under stress due to limitations on supply and high per capita consumption. Water rates tend to be higher under these conditions, and local officials face the real and recurring possibility of drinking water shortages. Water supply is less problematic in wetter, cooler climates, where groundwater and surface water resources are typically adequate to meet the local population's demands. Two climate variables appear in the model: annual mean total precipitation and mean daily maximum temperature, computed for the period from 1961 to 1990. These climate measures are not direct indicators of issue salience, but because they play a critical role in establishing the local supply of water resources and shaping patterns of demand, they should capture much of the cross-sectional variation across communities in the severity of water as a public problem. A local climate's vulnerability to drought conditions should be an apt proxy for the potential salience of water issues.

Results and Discussion

Results from this analysis demonstrate that governing structure has an important effect on the adoption of increasing block water rates, but the effect is conditional on local temperature conditions. Table 3.1 presents the results as marginal effects, or differences in the predicted probability of policy adoption associated with a shift from the twenty-fifth to the seventy-fifth percentile value of each independent variable—or from 0 to 1 for dichotomous variables—holding all other variables at their mean values. The table shows that special district governance has a positive effect on use of conservation rates in communities where maximum daily temperatures are low, but no impact at all in communities with hotter climates. Figure 3.2 shows the marginal effects over the range of maximum temperature values included in the dataset. At the median maximum daily temperature of 60 degrees Fahrenheit, the marginal

Table 3.1
Determinants of Adoption of Increasing Block Rates

Variable	Marginal effects
Special district (50 degrees maximum daily temperature)	.27***
Special district (70 degrees maximum daily temperature)	.01
Maximum daily temperature (among special districts)	.01
Maximum daily temperature (among cities and counties)	.26***
Mean annual precipitation	−.01
Operating ratio	.01
Percentage retail sales	.05***
Urban	.04
Surface water	.00
Purchased water	.11
West	.17*
Midwest	−.06
South	−.13

Notes: Cell entries show the difference in predicted probability of increasing block rate adoption associated with a shift from the twenty-fifth to the seventy-fifth percentile value of each independent variable (or from 0 to 1 for dichotomous variables), fixing all other variables at their mean values. Probabilities are based on estimates from the probit model that appears in the final column of table A1.2. Estimates are significant at $^*p < .10$, $^{**}p < .05$, $^{***}p < .01$ (two-tailed).

effect of specialization is 0.20, meaning that special districts are 20 percentage points more likely than cities and counties to employ residential rate structures that promote public good provision. The difference is larger where temperatures are low. In communities where the temperature reaches a maximum of 50 degrees on average, specialization produces a twenty-seven-point increase in the likelihood of a progressive rate structure. In higher temperature categories, however, the institutional effect on policy choice diminishes. Where temperature is at the seventy-fifth percentile level of 70 degrees, there is no significant difference across governance types in the likelihood of adopting conservation rates.[27]

Evaluating the direct effect of climate conditions on water rate structures, it is not surprising that increasing block rates are most popular among utilities located in the warmest regions. With utility governance held constant, water systems in regions with maximum temperatures at

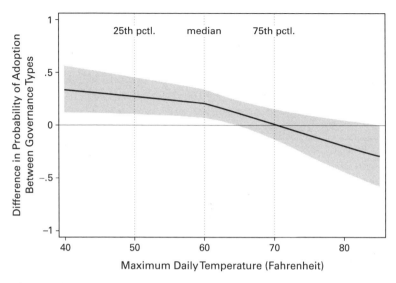

Figure 3.2
Effect of special district governance on adoption of increasing block rates. *Source:* Probit estimates in table A1.2. Gray band shows 95 percent confidence interval.

the seventy-fifth percentile are twice as likely to adopt conservation pricing as their counterparts with temperatures at the twenty-fifth percentile. As figure 3.3 demonstrates, the conditioning effect of temperature operates exclusively by changing general-purpose governments' policy choices. Shifting from the twenty-fifth to the seventy-fifth percentile temperature category produces a twenty-six-point increase in the likelihood that progressive rates will be used among utilities operated by a city or county; water districts' policy choices remain the same. Thus, the severity of a public problem influences how a multipurpose government balances competing demands associated with public good provision and the imposition of private costs. Where a problem is more serious, the public good benefits of policy action receive greater weight, and the general-purpose government becomes more responsive to the median resident's preferences. In contrast, special districts appear to make the same policy choices regardless of problem status.

The impact of problem severity is evident only in using maximum daily temperature as an indicator of climatic conditions. Total annual precipitation has no direct effect on use of increasing block rates, and nonlinear

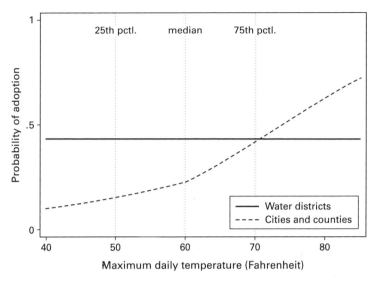

Figure 3.3
Probability of adoption of increasing block rates by governing structure. *Source:* Probit estimates in table A1.2. Lines show estimated probability of adoption, holding control variables at their mean.

and interactive relationships for precipitation also failed to pass standard hypothesis tests.[28] The absence of interaction between precipitation and governance type counters expectation. Given the strong findings for temperature, however, there is reason to interpret this null result as a weakness of the precipitation indicator for measuring problem severity rather than as evidence against the larger theory. Results in the water-demand literature reveal that, on the whole, temperature has a more powerful influence on water consumption than does precipitation. Moreover, there is greater variation in how that literature measures precipitation. Indicators include precipitation frequency, seasonality, and excess over evapotranspiration in addition to aggregate annual measures such as the one used here. These indicators imply different mechanisms for the relationship between precipitation and problem severity, and some even suggest that precipitation's effects are conditional on temperature.

Finally, temperature has a high degree of face validity as a measure of objective conditions regarding water. The scarcity of water supplies in the arid West is well known, but water has become an important issue throughout the Sunbelt. Groundwater depletion has produced subsidence in Houston and saltwater intrusion in Florida, prompting heightened attention to water issues even in these regions of abundant rainfall. Georgia's pursuit of new water storage to address growing resource scarcity has brought the state into prolonged conflict with Alabama and Florida. Drought conditions in the Southeast during the early 1990s and even more severely in 2007 have drawn attention to the vulnerability of the region's water supplies. In sum, the link between temperature and water scarcity is more apparent on its face and has been established with more certainty in the literature, and evidence of that stronger linkage is apparent in these results.

Control variables in the model contribute little to explaining policy choices regarding water rates. The one exception is the percentage of a utility's customer base that purchases water retail directly from the utility; an increase from 89 percent retail customers to 100 percent produces a five-point boost in likelihood of using increasing block rates. This result is consistent with what we would expect given the incentives for local officials, regardless of institutional setting or whether they are oriented toward policy or office goals. From a policy perspective, utilities that primarily serve retail customers should be more likely to invest time and resources to seek out efficient policies for retail users. Retail rates would be a lower priority for policy-oriented officials operating a primarily wholesale business. Office-seeking local officials also should respond to the composition of the customer base, because a larger retail population increases the potential salience of retail rates in an evaluation of local officials. Location in the West also increases the probability of using conservation rates, although the estimate of this effect lies just outside the 95 percent confidence interval.

These results suggest that the relationship between functional specialization and policy choice may be more complex than anticipated by both advocates and critics of special districts. Governing structure has an important influence on how public water utilities distribute the costs of service among residential users, but this influence varies with the severity of water scarcity as a public problem.[29] Throughout most of the United States, water is in adequate supply and so occupies little space on citizens and public officials' agenda. Dedicating a specialized venue to water issues allows decision makers to focus more attention on public good provision and the likely impacts of various rate structures on the median household's water bills. Water scarcity raises the stakes on water policy. It creates supply constraints and increases consumer demand for water, heightening the importance of conservation and efficient water distribution. Where high temperatures increase resource scarcity, water issues are more likely to capture public attention. Voters hold city and county officials accountable for their decisions on water policy, creating incentives for these officials to provide public benefits and a rate system that favors the median resident. When the public is attentive to water policy, the interests that stand to lose from increasing block structures do not enjoy the same advantage from monitoring and lobbying a multipurpose legislature as they do in a low-salience policy environment. Where problems are more severe, we do not see the same difference across institutional structures in the policy decisions they make.[30]

The Impact of Institutional Diversity among Special Districts

The analysis in the previous section demonstrates the conditional nature of the relationship between functional specialization and policy choices to provide public goods, but further investigation is needed to characterize the various institutional impacts that may arise from special district governance. Among special districts, rich structural diversity may play a role in guiding political action. Rules for participation in the selection of district officials define the political community and focus officeholders' attention on some interests over others. The existing literature on special districts gives substantial attention to institutional features but understates the diversity among district features. Critics of special districts point to exclusionary voting rules that restrict citizens' ability to express their preferences and hold district officials accountable for their decisions. This perspective ignores the districts that are highly visible to constituents and offer multiple avenues for electoral and civic participation. Conversely, those who support specialized governance often assume the existence of appropriate structures for democratic accountability, but these structures are absent in some special district arrangements. In order to understand the consequences of special district governance for representation, we must acknowledge that not all districts are equally representative.

Analyzing specific features of districts' institutional design is necessary to understand the practical implications of the rise of special district governance. It also can help identify the causal mechanism that accounts for differences in policy choices made by utilities operated by special districts and those managed by a city or county. The discussion thus far has assumed that outcomes can be attributed to the policy responsiveness of different institutional forms given a set of problem conditions. However, the important factor underlying the conditional effect of specialization may be issue expertise: specialization allows water districts to be more attentive to professional standards and policy innovations under normal conditions, but municipal governments can use consultants and other outside resources to close the gap in expertise when scarcity forces them to identify options for increasing supply or reducing demand.

The rules governing selection of special district officials provide some leverage for testing responsiveness as a causal mechanism. If political incentives explain the differences in policy choice between specialized and general-purpose governments, there also should be differences among special districts based on the political incentives created by districts' structural organization. Governments with election practices that minimize the bias in translating constituents' preferences into votes and seats should attract greater public participation and produce policies that are more consistent with public preferences. Institutions that introduce some bias to special districts' responsiveness—through selection of district officials by appointment rather than by election and elections by ward rather than at large—make it more likely that a special district will produce biased policy decisions. If a utility's rate policies respond to local officials' perceptions of constituent preferences, then selection rules that define a politician's constituency should have an influence on the likelihood of adopting a progressive rate.

Hypotheses

The analysis focuses first on the presence of special district elections. Notwithstanding critics' assertions that special districts lack mechanisms for electoral accountability, nearly 70 percent of water district officials nationwide are selected by a vote of district constituents. The rest are appointed to their positions by elected officials representing overlapping cities and counties or the state. Appointed officials should have less incentive to respond to constituents' preferences than elected officials have. Although the appointing authority for a district board is a set of elected officials who themselves should respond to electoral incentives in making water district appointments, the absence of elections for water district officials is, for several reasons, likely to introduce some distortion in the translation of constituents' preferences into policy.

First, the officials who make special district appointments often represent constituencies that differ from the special district's population. City and county officials may have appointing authority even if the special district encompasses just a small portion of their jurisdiction. In some cases, the state has appointing authority. The electoral incentives for these appointing officials will not necessarily correspond to the district's constituency. Moreover, with an appointed board, constituents' preferences on the selection of water district officials are bundled with their preferences on other issues overseen by the appointing authority (Besley and Coate 2003). Voters will cast a ballot for the appointing official based on issues that are more salient than water policy. When water district officials are directly elected, water rates might well be the most salient issue in the election. The incentives to respond to residents' preferences are stronger for elected officials. Finally, the transmission of constituent preferences to a special district board is simply more direct with an elected board; adding an appointing authority introduces friction that should be more likely to produce bias in outcomes.

The other institutional characteristic examined here is whether local government elections are organized by ward or at large. The formation of ward boundaries introduces the possibility of bias in aggregate representation, even when individual officials are purely responsive to their constituencies. In a jurisdiction that has been divided into wards, the median resident's position on increasing block rates will vary across wards based on patterns of wealth distribution and residential development. Ward boundaries are critical in aggregating constituent preferences, and their existence makes it more likely that a purely responsive legislature will produce a decision about the cross-subsidization of water costs that differs from the aggregate median resident's preference. Indeed, the adoption of at-large election structures at the local level was intended to promote politics that responded to majority preferences (Meier, Juenke, Wrinkle, et al. 2005). Numerous studies show that ward elections benefit the interests of minorities (Engstrom and McDonald 1981; Karnig and Welch 1982; Meier, Juenke, Wrinkle, et al. 2005; Welch 1990). In the case of water rate policy, wealthy homeowners whose rates would increase under progressive pricing make up the minority that stands to benefit from a biased system. At-large elections should allow for more direct translation of the median resident's preference on water rates. With an at-large system, all public officials represent the same constituency. If the governing board is truly responsive to residents' preferences, it will adopt a policy that satisfies the jurisdiction's median resident.[31]

As in the earlier analysis, the hypotheses here predict that institutional effects will vary with problem severity. Election of district officials and elections by ward will have their greatest effect where objective conditions are good and water has low potential salience. In communities where water supply is a more severe public problem, institutional design should be less relevant in shaping policy outcomes.

Results and Discussion

To evaluate whether responsiveness plays a role in explaining the difference between specialized and general-purpose governments in adoption of a progressive rate structure, I incorporated into the policy adoption model a set of institutional variables describing special district boundaries and the selection process for governing officials. The first hypothesis predicts that an elected board, rather than appointed special district officials, should be more likely to adopt progressive rates where problem severity is low. The results displayed in figure 3.4 demonstrate support for the hypothesis. Where mean maximum temperature is 60 degrees Fahrenheit, there is a twenty-five-point difference between elected and

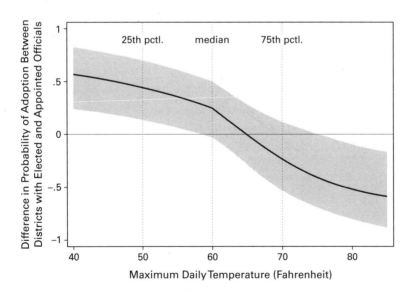

Figure 3.4
Effect of board elections on adoption of increasing block rates. *Source:* Probit estimates in the first column of table A1.1. Gray band shows 90 percent confidence interval.

appointed special district boards in likelihood of policy adoption. With the small sample of water districts in the analysis, this difference is not significant, but the institutional effect is larger and significant in climates colder than the median. In warm climates, the selection process for district officials has no significant effect on policy choice. Separating water from other dimensions of local policy by providing for direct election of water district officials seems to have the greatest impact where water might not factor into voting decisions otherwise.

Testing the hypotheses about ward elections involves two models: one for cities and counties that operate water utilities and one for water districts with elected board members.[32] Both models produce sizeable effect estimates for ward elections. Figure 3.5 shows that utility governing boards with a larger proportion of members elected by ward are significantly less likely to adopt progressive rates where water supply is plentiful. At the median temperature, the marginal effect of electing city and county council members by ward is to reduce use of a progressive rate by 11 percentage points. Among the small sample of elected water districts, the probabilities of using a progressive rate start higher, and the

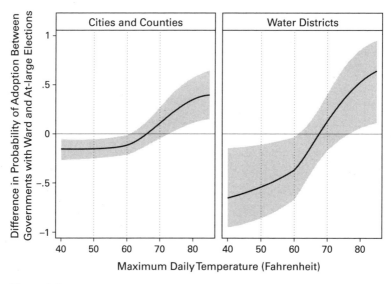

Figure 3.5
Effect of ward elections on adoption of increasing block rates. *Source:* Probit estimates in the second and third columns of table A1.1. Gray bands show 90 percent confidence intervals.

effect of election type is even more substantial: predicted probabilities for ward-elected and at-large boards are 0.40 and 0.76, respectively, so at-large elections almost double the likelihood of using a progressive rate. These differences across election types in median-temperature communities are significant at the 90 percent confidence level, as indicated by the gray bands. Here again, as hotter weather makes water supply a more important public problem, institutions have less influence on policy choice, although the relationship may reverse at the highest temperatures.

These results lend support to the hypothesis that variation in responsiveness accounts for the difference between special districts and general-purpose governments in their reliance on progressive rates. If responsiveness produces the outcome of policy adoption, then selection rules that promote responsiveness also should have an influence on adoption. Where residents have the opportunity to vote for officials who set water prices, and where votes are aggregated without the interference of politically driven ward boundaries that might distort the transmission of majority opinion on water pricing, officials should be more likely to choose rate structures that promote public goods such as water conservation. Private interests who bear the cost of providing the public good should have less influence under these conditions. The findings suggest that such is the case. Moreover, the results support the larger argument about the importance of policy context in conditioning the effects of institutional choice. When objective conditions in a policy area pose an important public problem, all politicians have an incentive to respond, and the organization of government matters less in determining policy outcomes.

Conclusion

For a century, American communities satisfied the public-health and drinking water needs of their growing populations by seeking out new water supplies and building large, centralized facilities for storage and treatment. The development of community water systems offered a safer and more reliable source of water to meet daily household needs. It also provided public goods that might not have been obtained otherwise, such as reduced risk from fire and infectious disease.

Communities continue to face water supply shortages caused by population growth, and their resources may be further strained by other factors, including aging infrastructure, environmental regulation, and climate change. But options for building their way out of these shortages have become limited. Groundwater aquifers are depleted, and surface sources have been dedicated to other uses. Higher hurdles to the construction of major facilities are also in place. The challenge facing water utilities now is to find a way to reduce demand so they no longer need to obtain new supplies.

Local water systems still have a role to play in providing public goods, but those goods now accrue from water conservation rather than from consumption. One tool utilities have for encouraging conservation is increasing block rates, which offer the added public benefits of economic efficiency and vertical equity. The shift from private to public governance of water utilities in the nineteenth century had an important impact on the provision of public goods. This chapter has examined how the more recent trend toward specialization of public utilities has affected the provision of public goods in the current era.

The findings suggest that governance factors still matter. Under normal circumstances, specialized governance of water utilities makes it more likely that utilities will make policy choices that advance equity and resource conservation. These choices appear to be explained not by the issue expertise and greater dedication to professional standards that characterize specialized bodies, but rather by differences in responsiveness to constituents' preferences. The gap between special districts and traditional local governments with regard to responsiveness is not universal, however. Special districts that are less visible to the public, such as those with governing-board members that are appointed rather than elected, make policy choices that demonstrate greater bias toward private interests. And in hot climates where water is scarce and the possibility of high rates and shortages raises the salience of water issues, institutional factors matter less. All governments confronting a serious problem have an incentive to provide public goods that might help solve the problem.

4

Distributing the Price of Growth

One strategy for reducing demands on a strained water system is to promote conservation among existing customers. Another strategy is to discourage new connections that expand the customer base and shrink the surplus water supply that most utilities reserve for periods of drought. Utilities can discourage new connections either by mandate or through incentives. Systems facing severe shortages may declare a moratorium on new connections, until either drought conditions ease or the utility can secure new supply or expand its infrastructure. Even where overall water supply is plentiful, utilities sometimes adopt moratoria because of capacity limits in a utility's storage, treatment, and distribution facilities.

Restrictions on new water connections or overall building construction are an effective instrument for halting the rise in demands on a water system. Cutting off growth is a severe policy act, however, involving high political costs and potential legal liability. An incentive-based strategy designed to achieve the same goal is to impose impact fees on new residential water connections. Part of the larger category of development fees and land-use exactions, impact fees are a charge to developers (or, less commonly, to existing homeowners) for a new connection to a water system. Developers pass the fee on to home buyers as part of the cost of new housing, and water utilities use the revenue to pay for system expansion. The effect is to shift the cost of expansion to new customers rather than divide it among the entire customer base. Impact fees are a water-pricing policy, but even more important a growth policy. By increasing costs to developers and new homeowners, proponents expect impact fees to provide a disincentive for growth and the ensuing demands on a community's water infrastructure. Many communities employ the fees as part of a long-range plan to promote sustainable management of local water resources.

This chapter examines California water utilities' growing reliance on impact fees to fund the cost of water-system expansion. It investigates the determinants of impact fee levels, focusing on the influence of utility governing structure. Impact fees are a pricing tool that lies at the intersection of water and land-use policy. Where water and land use fall under the jurisdiction of separate local governments, a water district's reliance on impact fees can be seen as an encroachment on the overlapping city or county's authority, with or without that government's consent. We might expect, therefore, that fragmentation of authority would suppress the use of impact fees, especially if water districts are beholden to development interests, as the conventional wisdom suggests. In fact, the analysis demonstrates that in many contexts, water districts charge higher connection fees than do cities and counties that operate water utilities. Despite formal limitations on their responsibilities, water districts are willing to employ pricing strategies that have important implications for land-use policy. The effect of specialization is conditional on the growth rate within a community, however, and where development pressures are high, governing structure has little influence over impact fee levels. These results suggest an expansion of growth conflicts into water district venues where the politics of growth are most contentious.

The Rise of Development Fees as a Local Revenue Source

Water impact fees are a response to the pressure that population growth places on a community's water supply infrastructure. Historically, water systems were built using general revenues or bond financing, and the entire customer base shared in paying the cost of system construction and expansion. Existing residents subsidized water service for newcomers to a community, just as newcomers contributed to the maintenance and repair of facilities whose condition may have deteriorated before they arrived. As Alan Altshuler and José Gómez-Ibáñez (1993) describe in their comprehensive study of development fees, this arrangement reflected the dominant belief that the economic benefits of population growth outweighed the costs of providing public services for new residents. The recent popularity of impact fees reflects a shift away from this belief. As more people come to believe that development does not pay its own way, they expect new residents to bear a larger share of the cost of expanding infrastructure to meet rising water demand.

Water impact fees are just one example of a larger effort to shift the costs of growth to new residents.[1] Communities throughout the nation are increasingly imposing development fees and land-use exactions on developers seeking project approval. Altshuler and Gómez-Ibáñez report that the share of communities charging land-use exactions rose from 10 percent in 1960 to 90 percent in the mid-1980s (1993, 125). Some of these exactions are in the form of in-kind goods and services, such as the construction of public facilities or the dedication of public land. Fees are the most common form, however, and a 2000 survey conducted by the U.S. Government Accountability Office (GAO, previously the General Accounting Office) showed that more than half of cities and counties with a population higher than twenty-five thousand imposed monetary impact fees. California is highly reliant on this revenue source: more than 90 percent of local governments surveyed by the GAO required impact fees from developers (U.S. GAO 2000). A 1999 survey revealed that average fees for a new single-family home in a sample of eighty-nine California cities and counties totaled $24,325 (Landis, Larice, Dawson, et al. 1999). A few communities have treated exactions as a means for developers to subsidize a public good the community otherwise cannot provide, and thus much of the legal attention to exactions has focused on the Supreme Court's requirement that a nexus exist between the exaction and the state interest being advanced by the mandate.[2]

Like user fees, exactions reflect an effort to privatize the cost of public goods and services. This effort has become necessary due to the obstacles that local governments face in raising the revenue required to update the nation's aging infrastructure. State limitations on local tax and debt authority and increased public resistance to new taxes make it difficult for localities to generate revenue from their own sources, and cutbacks at the state and federal levels have reduced the number of alternative financing options available.[3]

Most important in explaining the rise of impact fees is the increased community resistance to growth that has emerged in recent decades. As the public has become more concerned about social and environmental costs of new development and more skeptical about its economic benefits, governments have sought ways to slow down growth and shift its costs away from existing residents. Local officials view impact fees as a means to achieve both goals. The fees heighten the cost of residential development, sometimes adding tens of thousands of dollars per housing unit, thereby increasing the developer's investment risk and raising home prices for new residents. Impact fees also exempt existing residents from bearing the cost of expanding facilities and services to meet growing demand on public services. Impact fees are likely to receive broad support from citizens for reducing the public burden of financing infrastructure costs and strong backing from environmentalists and antigrowth interests for providing a disincentive to further development. As Altshuler and Gómez-Ibáñez describe, the fees offer an attractive policy solution for local officials seeking to accommodate new development without losing the support of their constituents: "By obtaining developer commitments to finance public facilities and services, local officials can maintain that they have protected the interest of current residents and, more generally, that they are 'managing' growth rather than caving in to either developers or antigrowth extremists" (1993, 47). By transferring costs to future residents, fees make local decisions to approve new development somewhat more tolerable to those who oppose growth.[4]

Although impact fees receive broad public support, they are likely to draw intense opposition from developers and large landowners. Who bears the cost of impact fees depends on local conditions in the housing market, but developers recognize the risk that they will absorb some part of the fee if the market prohibits them from passing it along fully as part of home prices. Unanticipated changes in impact fee levels might affect developers' profits in the short run regardless of market conditions. If impact fees reduce returns on development investments, developers will lower their bids on land for new construction, diminishing landowners' profits. Given the likely burden of costs, it is not surprising that proposals for new or increased impact fees have often drawn strong resistance from both of these groups.[5]

Water Impact Fees as a Form of Growth Policy

Water impact fees are one of the most common forms of land-use exaction (Been 2005; Landis, Larice, Dawson, et al. 1999), and, as with other fees, their use has increased over time (Weschler, Mushkatel, and Frank 1987). Water impact fees differ from many other forms of land-use exaction in that the nexus between the fee and its purpose is apparent. Developers pay a fee for the right to attach a new line to the system; the revenues cover the cost of infrastructure needed to provide service to the new development. Impact fees are explicitly a mechanism for funding water-system expansion—not for distributing the cost of system operations—and they reflect a policy position that development should pay its own way. As a result, although the nexus is clear between the revenue source and its application, the impact of the fees reaches beyond water policy to affect land-use planning as well.

There are several reasons why analysis of impact fees is relevant for understanding the new local politics of water. First, impact fees are an important potential source of revenue to meet the nation's large and growing infrastructure investment needs. The U.S. EPA (2005) has estimated that community water systems will have to invest $270 billion in infrastructure expansion and improvements over the next twenty years in order to provide safe and clean drinking water to their customers. This estimate does not include projects that would be undertaken solely to accommodate future growth. Given local preferences to keep water rates low and limitations on state and federal assistance, the agency has predicted the possibility of a significant funding gap between needs and spending (U.S. EPA 2002b). Impact fees are one strategy for filling the gap that does not require across-the-board escalation in water rates. They can be used to replace dilapidated infrastructure with new facilities that meet the demands of a growing community. They also can fund the infrastructure expansion needs not included in the EPA's gap analysis, such as extension of water pipes to new housing developments. With most local revenue sources in decline, it is critical that we examine one of the few options available for generating new revenue and understand the determinants of its use.

Second, by existing at the intersection of water and land-use policy, impact fees can be a strategy to incorporate consideration of water adequacy into decision making about land use. A demand-centered approach to water supply management requires coordination of water and land-use planning. Adding new connections to a water system that lacks adequate supply or facilities can jeopardize service to existing customers and deplete the supply source. Development may introduce contaminants into the water supply and reduce groundwater recharge, and it can lock in long-term patterns of water demand. Strategies for meeting these planning challenges sometimes take the form of mandates, as in the case of local moratoria on new connections or state requirements that adequacy of water supply be demonstrated before approval of new development.[6] Impact fees are a milder, incentive-based approach intended to slow the rise in demands on water systems that operate near capacity. Antigrowth interests may also endorse impact fees as a means to hold up new development even where supply and infrastructure are adequate to meet the demands of a growing community. In either case, impact fees are evidence of the increasing interdependence between water and land-use planning and of the ever more important role of local governments in solving the nation's water problems.

Third, analysis of impact fee use can help reveal the dynamics of specialized governance. The growing need for coordinated planning coincides with increased fragmentation of local governing authority. Setting impact fee levels falls under the jurisdiction of a water utility, but the policy decision has important consequences for a community's ability to achieve its growth goals. Does dividing responsibility for water and land use into separate jurisdictions reduce a water utility's reliance on this policy tool? Private water companies typically do not charge impact fees. Water districts, too, may avoid them out of respect for city and county land-use authorities or because they are captured by development interests that oppose the fees. Districts might alternatively impose high impact fees as a direct challenge to the local land-use authority's policies. Separation of responsibility does not rule out coordination, however, and water districts may cooperate with overlapping general-purpose jurisdictions to pursue shared goals regarding the management of water and land use.

The analysis in the next section considers the effects of special district governance on water impact fee reliance. The academic literature on impact fees has focused largely on their effects, examining the impact on housing and land prices (Dresch and Sheffrin 1997; Evans-Cowley, Forgey, and Rutherford 2005; Ihlanfeldt and Shaughnessy 2004; Singell and Lillydahl 1990; Yinger 1998), housing supply (Brueckner 1997; Hanak 2008; Hanak and Chen 2007; Mayer and Somerville 2000), capital investment (Clarke and Evans 1999), and job growth (Jeong and Feiock 2006). The determinants of impact fee reliance have received less attention, especially in the area of water impact fees. A 1985 national survey of cities and counties showed a strong positive relationship between a community's growth rate and the likelihood of impact fee adoption (Purdum and Frank 1987). A 1986 survey conducted in nine southeastern states found that water and sewer impact fee adoption was a function of local demand and the government's capacity to innovate and coordinate water and land-use planning (Kaiser, Burby, and Moreau 1988). The most recent and detailed analysis examines impact fee adoption in sixty-six Florida counties over twenty-five years for all functions except water and sewers (Jeong 2006). Important factors that emerged in this study are population growth, which promotes impact fee use, and the strength of the development community, which has a negative effect on fee use.

All of these previous studies have focused on impact fee adoption, ignoring the steady rise in fee levels over time. Local governments might adopt impact fees as a policy innovation or to meet a short-term revenue shortfall without substantially shifting the burden of paying for infrastructure improvements. Factors contributing to the decision to create an impact fee might not explain the marked growth in fee levels during the 1990s. Nominal fees should have little effect on existing residents' obligation to pay for growth or on developers' calculations about plans for new construction. They also might not attract much opposition if developers have confidence that the fees will stay low. High fees reveal a commitment to shifting the costs of growth to new residents. Moreover, previous surveys of impact fee usage have focused on cities and counties only; we know nothing about the diffusion of this policy tool among special districts.[7] The analysis presented in this chapter aims to explain the increased reliance on impact fees, paying particular attention to how different government types use this revenue source. It shows that governmental structure interacts with local conditions to influence how a community chooses to finance growth.

Reliance on Impact Fees

Like the increasing block rate structures discussed in chapter 3, impact fees have important distributional consequences in that they allocate the burden of paying for water-system expansion. Unlike block rate structures, however, impact fees offer few public good benefits. As a fixed cost, they do not send market signals about water supply that help encourage conservation. They rarely reflect the marginal cost of system expansion, so they fail to promote economic efficiency.[8] Impact fees also tend to be regressive. They typically impose the same fee regardless of home price, thus contributing to a larger increase in housing costs for those who spend less on housing.[9] Assuming relatively recent adoption, they transfer costs from existing to future residents and from older to younger generations, which may have an additional regressive effect in terms of income redistribution (Altshuler and Gómez-Ibáñez 1993).

Without public good benefits, the significance of fee levels directly relates to how the costs of growth get distributed. Water utilities can fund infrastructure expansion either with impact fees or through taxes, bonds, or user fees that distribute costs across the entire customer base. Reliance on impact fees reflects an explicit decision to make growth pay its own way, typically in the face of opposition from developers and large landowners. It reveals a preference to shift the cost burden of new development to incoming residents and perhaps to slow down the pace of growth. Consequently, analysis of impact fee levels can reveal the conditions under which water districts extend their scope to consider questions about growth and how they respond to the interests positioned on either side of a growth debate.

Hypotheses

Planning for growth is a function that belongs to cities and counties, but it requires the cooperation of special districts that provide services to the areas proposed for new development. Special districts are an important part of the battleground over growth in communities throughout the United States. They can delay construction on specific projects by refusing to provide public services, or they can promote growth by extending services to previously undeveloped areas. A district may be formally responsible only for providing water, sewers, or roads, but its decisions about where to extend those services have inescapable implications for land development. Moreover, in deciding how to distribute the cost of expanding infrastructure to serve new residents, special districts may help or hinder a community's efforts to control and direct its growth.

Despite the myriad interrelationships between service-delivery decisions and growth, it is rare for special district officials to acknowledge or embrace their role in land-use planning. In public statements and campaign literature, special district officials consistently defer to city and county authority over growth decisions. But district officials have both opportunity and motive to exert independent influence over land-use policy; they may have private preferences about growth and development that depart from the local land-use authority's approach, or they may seek to build support for future political office through policy choices that have implications for salient local growth disputes. Special districts also are likely to receive more attention and lobbying from residents and interest groups when their decisions touch on growth issues.

The establishment of impact fees is one means by which water districts might influence land-use outcomes, either with or without the overlapping land-use authority's endorsement. In a functionally fragmented system, special district cooperation is necessary if a city or county wants to use pricing strategies as a mechanism for slowing down growth. The general-purpose government that has jurisdiction over land use does not control the facilities required to serve new development. It must persuade the relevant special districts to adopt policies that support local growth goals.

If the conventional wisdom about specialized governance is correct, special districts should be disinclined to cooperate in this case. Narrowly defined special districts are not designed to respond to citizens' preferences about growth, and their prodevelopment bias should make them unlikely to embrace the use of impact fees. Critics argue that the special districts' single-issue focus leads to fractured decision making, which inhibits the kind of policy coordination that impact fees might represent. In a classic work on special districts, John Bollens argues:

[The] piecemeal, unintelligent attack on the problems of government, and the lack of over-all administrative and policy planning which grows out of the proliferation of governmental units, hinder the orderly development and sound utilization of the resources of an area. The approaches of different governments to a common problem often conflict and work at cross purposes, thus dissipating needed energies. A special district that handles only one aspect of a many-sided problem may do so with harmful results. (1957, 255)

Victor Jones, another leading metropolitan reformer, predicted that proliferation of special districts would result in "the further disintegration of authority and dispersion of control, the increase of ruinous competition for available tax resources, and continued uncoordinated planning of governmental services" in a metropolitan region (1942, xxi). This view suggests that special districts will be disinclined to consider the effects of their decisions on issues of growth and development or to cooperate with neighboring governments on a comprehensive growth strategy.[10]

Special districts also might limit their use of impact fees if they are biased in favor of promoting growth, as many analysts suggest. Developers frequently play an important role in establishing special districts as a means to finance public facilities in areas that existing municipalities cannot or will not serve. Nancy Burns (1994) has argued that developers institutionalize their preferences when they establish a new district, creating a policy venue in which the progrowth agenda rarely faces challenge. Case studies of special district operations have provided some evidence that special districts are vulnerable to capture by developers and landowners, providing infrastructure to support growth even in the face of resource scarcity and financial risk. If special districts indeed demonstrate a progrowth bias, we should not see them choosing to impose high impact fees, a policy instrument that developers strongly oppose.

It is possible, however, that special districts might be more reliant on impact fees than a water utility operated by a city or county government is. Although some researchers have found special districts to be hospitable venues for developers and progrowth interests, it is cities that should see a greater benefit in promoting growth. Municipalities traditionally have perceived that attracting new residents—especially high-income residents—can stimulate economic development and expand the community's tax base. This attitude is becoming less dominant as more cities start to view the public costs of growth as outweighing its benefits, but cities still should see greater advantage in expanding the tax base than do special districts, and therefore cities may be less likely to adopt a policy perceived to create disincentives for new development. This hypothesis is consistent with the public choice theory of specialized governance: if large, consolidated governments provide an institutional advantage to interests with resources to invest in lobbying, then developers should be more successful in their efforts to fight impact fees in city and county venues. Special districts should be more responsive to majoritarian preferences to shift the cost of development to future residents. Moreover, if existing residents perceive that city government is dominated by a growth machine, those who seek to slow the pace of development may attempt to exploit policy opportunities in alternative venues such as special districts.

The baseline for comparison is a municipal water department. A consolidated city or county government that seeks to shift the cost of growth to new residents would direct its relevant departments to increase the impact fees for parks, roads, water and sewer infrastructure, and other public facilities. Thus, water districts that cooperate with overlapping general-purpose jurisdictions on local land-use goals should be equally reliant on impact fees as city water departments, all else held constant. If specialization has no effect, we can assume either shared goals or a cooperative relationship. Higher impact fees among cities would indicate the absence of cooperation between specialized and general-purpose jurisdictions in planning for growth, perhaps due to a prodevelopment bias on the part of special districts. Alternatively, higher impact fee use among water districts would be evidence against this bias and would suggest that special districts are acting outside the boundaries of their functional jurisdictions and becoming alternative venues for growth disputes.

In addition to these hypotheses about specialization's direct effect on impact fee reliance, I also consider whether the impact of governing structure is conditional on the status of growth as an issue in the community. The previous chapter revealed that differences in water rate choices made by specialized and general-purpose governments depend on the severity of water conditions. Like the issue of increasing block rates, the debate over impact fees typically positions a diffuse majority in favor of the policy against a smaller group that adamantly opposes it and has more resources to spend. In the case of impact fees, the policy issue at stake is growth, not water. The overall context of land development in a community may affect different government structures' responsiveness to progrowth and antigrowth arguments and the likelihood that interest groups will seek to expand conflicts over growth into new venues. The analysis tests this conditional hypothesis against competing predictions that the effects of governing structure are constant.

Model

The model examines impact fee levels among a sample of California public water utilities at two-year intervals during the period from 1991 to 2003.[11] Focusing on California allows me to control for state-level legal and regulatory factors that might affect water-pricing decisions, and it provides the opportunity to include control variables in the analysis that are not measurable when using a nationwide dataset.[12] Conducting the analysis only on utilities located in California has some disadvantages for external validity. Objective conditions related to water are more severe throughout California than they are in many other states. Indeed, more than three-quarters of the utilities included in this analysis are located in areas with temperatures that are higher than the median in the national dataset examined in the previous chapter. Moreover, growth rates are higher in California than they are in the rest of the nation, although the difference is modest and there exists substantial intrastate variation in the severity of growth issues.[13] Most of the state's coastal communities have effectively stalled new development, but many cities in the Central Valley and Inland Empire continue to experience high rates of population growth.

Another reason that findings from California may not generalize to other regions is that water impact fee levels might be higher in California than elsewhere. National surveys indicate that California communities are most likely to adopt land-use exactions due to limitations on use of other local revenue sources. In 1978, California citizens passed Proposition 13, which capped property tax rates statewide. Within a year of Prop 13's passage, a survey conducted by the California Building Industry Association revealed that the median bill for construction-related fees had risen 26 percent (Frieden 1983). Respondents to a contemporaneous survey conducted by the Association of Bay Area Governments reported that their fees had doubled or even tripled during that time (Landis, Larice, Dawson, et al. 1999, 21). By requiring voter approval for all taxes and most fees that are "property related," Proposition 218, passed in 1996, eliminated many of the fees that local governments had adopted as alternative revenue sources after Prop 13.

These restrictions on alternative revenue sources contribute to overall heavy reliance on impact fees among California's local governments, but water utilities in particular have the same options available for funding infrastructure expansion that they have in other states. Although Prop 218 limited the use of some impact fees, utilities still can impose water impact fees without voter approval.[14] And after passage of Prop 13, the state established an alternative instrument for spreading the costs of infrastructure expansion among all utility customers with voter approval.[15] One important feature of this analysis is that California's unique set of rules regarding local fees and taxes affect both cities and special districts, so the rules should not influence estimates of the impact of specialization on decisions about how to distribute the costs of growth.[16]

As in the analysis of utility rate structures, the key explanatory variables are governing structure and measures of problem conditions. Here the central policy problem of interest is growth. The dynamics of growth politics may vary based on the level of growth in the community. Where population is rapidly increasing, existing residents become concerned about the effects of growth on their quality of life, and they may resist subsidizing the expansion of public facilities to provide for new residents. Providing disincentives for new development is less relevant in communities experiencing low levels of growth, even if growth itself is a contentious issue. Growth rates also reveal the level of developer activity in the community. In areas of rapid growth, developers have an ongoing presence in the community and will likely learn more about the configuration of governments that oversee service delivery. They obtain a better understanding about the importance of decisions made by low-profile special districts in shaping the context for new development. In this model, the measure of growth is the rate of population increase in the jurisdiction between 1990 and 2000.

The model assumes that growth affects impact fee levels, but not the reverse. This assumption counters many fee advocates' expectation that impact fees serve as a deterrent to new housing development. In fact, recent economic analyses demonstrate that water impact fees do not slow down growth (Hanak 2008; Hanak and Chen 2007; Mayer and Somerville 2000). One of these studies is particularly relevant because it focuses on California utilities and accounts for changes in fee levels in addition to the existence of water impact fees. Ellen Hanak (2008) finds that unlike various other water-screening policies, neither fee adoption nor a change in fees has a significant effect on housing growth.

Also included in the model are measures of climate conditions. These conditions exhibit limited variation within California and therefore are less likely to be important determinants of pricing policy in a state-specific sample. Moreover, the state's reliance on large, surface water projects may create incongruity between local conditions and availability of supply. Nonetheless, measures are included to capture differences in temperature and rainfall both cross-sectionally and over time. In 1991, California was in the final year of a four-year drought that raised public awareness about water scarcity and the fragility of the state's long-term water supply. Public attention focused on water issues, as many utilities adopted rate increases and voluntary conservation programs in response to shortages. Some utilities even imposed mandatory usage restrictions. As California emerged from drought in the 1990s and began to enjoy a booming economy, public interest in water issues waned. The analysis tests whether special districts responded differently than city and county utilities to these changes in water salience. Time-variant measures of climate indicate annual departures from mean temperature and precipitation, with a two-year lag in order to provide time for a policy response. Also included is a statewide drought index, modeled with a two-year lag.

Control variables fall into two general categories: the constituency's demographic and political features that might prompt demand for growth control, and the utility's financial and operations characteristics. Because antigrowth efforts have often been characterized as movements by wealthy homeowners to protect their home values and create exclusionary communities, I include in the model the median income in the utility's jurisdiction. Urban, liberal populations also are perceived to be more likely to demand growth control. Measures of jurisdiction location and political ideology account for these factors. Also included in the analysis is the proportion of the jurisdiction's housing units that are in buildings with multiple residences. Many residents of multiunit buildings do not pay their own water bills and would likely have little reason to favor higher impact fees, thus dissipating constituent demand.

On the utility side, the model controls first for the level of a utility's water prices, because utilities that already charge high water rates might have more incentive to shift the cost of infrastructure expansion away from existing customers. A utility with large debt obligations also may be more likely to use impact fees as an alternative to issuing more bonds, so a measure of the amount of debt per capita attributable to the government's water-provision activities also appears. A measure of the city or special district's population captures the effect that economies of scale might have on the utility's financial decisions. Finally, the utility's supply source is included to account for its influence on the cost of water provision.

Results

Analysis of simple bivariate relationships reveals that water impact fees are not just a tool used by cities and counties that have authority over both water and land use. Figure 4.1 shows average household impact fee levels by year for the special districts and general-purpose governments in the sample. Not only do special districts use water impact fees, they rely on them more heavily than do their general-purpose counterparts. The differences in means between governing structures are significant at the $p < .01$ level in every year. The figure shows a drop-off in fees in 1993 among both utility types as water systems began to recover from the effects of California's extended drought. Fees rose sharply in 1995 and have risen gradually since that time to a level surpassing average fees in 1991.[17]

Figure 4.2 displays results from a panel data analysis that controls for other factors potentially affecting reliance on impact fees.[18] The figure shows the estimated dollar value of water impact fees per household imposed by specialized and general-purpose governments across the full range of growth conditions that exist in the dataset, holding control variables constant at their mean values. As the figure demonstrates, a high growth rate has opposite effects on the two governance types: it increases

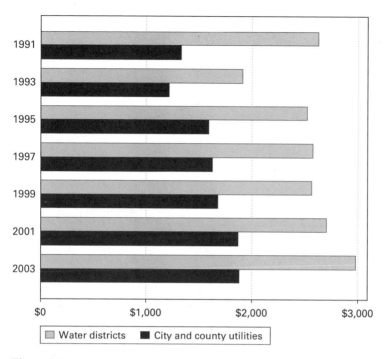

Figure 4.1
Average per household impact fees charged by California water utilities, 1991–2003. *Source:* Black & Veatch California Water Charge Surveys, 1991–2003.

the impact fees levied by municipal utilities, but decreases the fees that water districts impose. In most contexts, estimated impact fees are higher among special districts than among their general-purpose counterparts. In communities with the highest growth rates, the effect of governing structures reverses direction. The difference between the estimated fee levels appears in figure 4.3, along with a confidence band showing whether the effect of governing structure is statistically significant at different levels of population growth. Where population growth is modest, water districts impose higher fees. In communities with the twenty-fifth percentile level of 7 percent population growth between 1990 and 2000, specialized governance is predicted to increase impact fees by more than $500. This difference is larger where growth is even slower, and it shrinks at higher growth rates. For a community with the median 14 percent rate of growth, impact fees are an estimated $400 higher if water is overseen by a special district.[19] This difference is significant at the

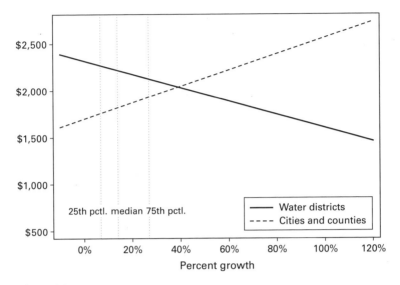

Figure 4.2
Estimated per household water impact fees by governing structure. *Source:* Prais-Winsten estimates in table A2.1. Lines show mean estimated impact fees, holding control variables at their mean.

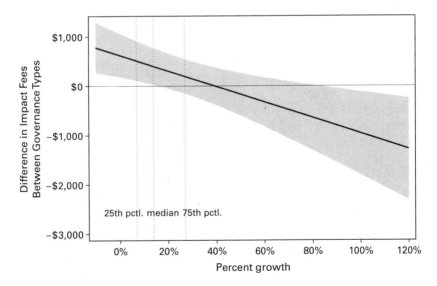

Figure 4.3
Effect of special district governance on water impact fee levels. *Source:* Prais-Winsten estimates in table A2.1. Gray band shows 90 percent confidence interval.

90 percent confidence level. We cannot say with confidence that governing structure has any effect on fee levels in faster-growing regions, where growth is between the median and the ninety-fifth percentile, but in the few communities that are booming most, special districts are predicted to charge less than cities and counties that operate water utilities.

Previous literature on impact fees has shown a positive relationship between growth and the existence of impact fees among cities and counties, but this analysis provides little evidence that higher growth rates influence a general-purpose government's reliance on fees. The effect of growth can be seen in figure 4.2 as well as in table 4.1, which shows the marginal effect of change in each independent variable from its twenty-fifth to its seventy-fifth percentile value—or from 0 to 1 for the two dichotomous variables—on the total dollar level of a water impact fee. A shift from 7 percent to 27 percent growth over a decade is estimated to increase fee levels by $173. The same shift has a somewhat smaller negative effect among special districts, reducing the fees that they impose by $136. Neither of these effects can be distinguished from zero with a high degree of confidence, and it is only in a shift from its lowest to its highest possible value that we see growth exhibiting a significant effect on fee levels.

Although these results are modest in strength, they lend support to the hypothesis that the effects of governing structure are conditional on the status of growth issues in the community. Both types of government appear to respond to real conditions of growth in their decision making about how to distribute its costs. Their responses are contrary to one another, however, with the consequence that specialization's effects vary according to a community's development pressures. When growth is at or below California's median rate, special districts set fee levels significantly higher than what cities and counties choose under the same conditions. Districts are more likely to rely on this policy tool where developers are not politically active and growth is not putting heavy strain on existing resources. Impact fees generate little revenue where growth rates are low; therefore, it is likely that special districts are setting the fees in response to community preferences rather than to financial considerations. Water districts appear not to define their responsibilities so narrowly that they ignore demands for policy action that might spill over issue boundaries. Cities and counties charge lower impact fees in these communities growing at a modest rate. This decision may be intentional in order to stimulate development, or it may be another

Table 4.1
Determinants of Impact Fee Levels

Variable	Effect on impact fee level
Special district (in communities with 7 percent growth, 1990–2000)	$510*
Special district (in communities with 27 percent growth, 1990–2000)	$194
Growth (among special districts)	−$136
Growth (among cities and counties)	$173
Climate variables:	
Average maximum daily temperature	$85
Average annual precipitation	$393**
Departure from normal daily temperature, two-year lag	$99*
Departure from normal precipitation, two-year lag	−$53
Drought index, two-year lag	$44
Demand variables:	
Median income	$448***
Proportion multiunit housing	−$398***
Proportion urban	−$54
Proportion Democrat	−$74
Utility variables:	
Water charge	$285***
Water debt per capita	$57*
Population served	$301***
Reliance on surface water	$521***

Notes: Cell entries show the difference in impact fees associated with a shift from the twenty-fifth to the seventy-fifth percentile value of each independent variable (or from 0 to 1 for the two dichotomous variables, special district and surface-water reliance). Effects are based on estimates from a Prais-Winsten panel data model that appears in table A2.1. Estimates are significant at *$p < .10$, **$p < .05$, ***$p < .01$ (two-tailed).

demonstration of policy inaction in a general-purpose venue when problem severity is low.

Where growth rates are higher than average, governing structure matters less. In this case, special districts appear to cooperate with their neighbors' land-use goals. Separating decision making for water from decision making for land use has little discernable impact on the use of pricing policies to distribute the costs of growth. The convergence is attributable to both general-purpose and specialized governments' response to higher growth rates. Where development exerts pressure on a community's resources, cities and counties set fees at a higher level. Constituents' demands to redistribute the costs of growth become more forceful and balance developers' opposition to these fees. At the same time, special districts attract more attention from developers where growth rates are high and become important venues for growth politics. High levels of growth increase the risk that special districts might deny service to new development because of limited system capacity or in response to constituent backlash against growth. In order to secure cooperation with service provision, developers invest more resources in lobbying special districts, which then better positions the developers to prevent district actions to increase impact fees. These efforts balance constituents' more diffuse calls for higher fees. Districts in rapidly growing communities also may be more hesitant to make use of a policy instrument that encroaches on the city or county land-use authority. In sum, the dynamics of specialized governance and interest groups' response to the incentives it creates can account for these patterns of policy choice.[20]

Climate conditions appear to have little effect on water impact fees.[21] Weather that is hotter than normal for a region pushes fee levels upward, but water agencies do not appear to respond to annual fluctuations in drought conditions by raising impact fees.[22] Among the cross-sectional climate variables, precipitation has a surprisingly large and significant positive effect on impact fee levels. It is likely that this effect is attributable not to actual precipitation levels, but rather to the important cultural differences between California's coastal and inland communities, the latter of which receive less rainfall and generally have demonstrated less resistance to growth.

Among control variables, characteristics of both the jurisdiction and the utility have an impact on fee levels. Average median income has a sizeable impact: a shift from the twenty-fifth to seventy-fifth percentile in income increases estimated fees by nearly $450. The proportion of a jurisdiction's homes that are in multiunit buildings has a negative effect by dissipating constituent demands for impact fees. The larger the proportion of residents who do not pay their water bills directly, the less demand there exists for fee structures that shift costs away from existing customers. The results provide no support for hypotheses that impact fees will be higher in liberal, urban communities. All the utility variables have significant effects. Impact fees are higher among large utilities, and there is evidence that water systems use impact fees as a means to overcome financial constraints. Fee levels are positively associated with reliance on surface water, per capita water debt, and the level of residential water bills.

Conclusion

Impact fees have become a point of conflict in local disputes over growth. Environmental and community groups often seek high impact fees for their expected disincentive effects on development and the ability to impose the costs of growth on future residents. Builders and landowners try to keep impact fees low to protect housing affordability and their own profit margins. Although water impact fees can be a means for distributing water supplies more effectively and addressing revenue shortfalls, they are more commonly framed as a growth-control strategy. The fee-setting process then becomes part of a larger local battle over land-use policy.

Previous analyses of impact fees have focused on their use by general-purpose local governments as part of a comprehensive strategy for fiscal management and land use. Impact fees are not just a tool available to cities and counties, however. Special districts can levy fees as well to finance the facilities required by a growing population. The separation of responsibility for water use from responsibility for land use poses a dilemma for reliance on impact fees as a policy instrument. Water districts that impose the fees risk encroaching on the land-use authority of overlapping cities and counties. Moreover, critics of specialized governance worry that fragmentation of authority over issues related to growth planning may create opportunity for developers to dominate the policy process. We might expect then that specialization would lessen reliance on impact fees, because the fees have effects that cross over functional boundaries and attract opposition from developers.

The analysis in this chapter reveals that these expectations do not hold true. Water districts are at least as reliant on impact fees as cities and counties are, and in slow-growing communities they impose higher fees than general-purpose governments do. Special district officials appear willing to use their oversight of local services in order to pursue growth-management goals and sometimes even take policy actions that developers firmly oppose. This does not mean that districts are invulnerable to pressure from special interests. Districts show the most willingness to use this tool where developers are least likely to be paying attention.

Where growth rates are low and developers restrict their attention to the city and county governments responsible for land use, special districts impose the highest impact fees, and governing structure has the largest effect on impact fee levels. Districts act outside their area of functional jurisdiction in order to respond to calls from their constituents to shift the cost of growth to incoming residents. Cities and counties are less attentive to these diffuse demands for impact fees, especially when they are countered by opposition from the development community.

With higher growth rates, a community becomes more profitable and attractive to developers. Builders and landowners invest more resources in lobbying all the governments responsible for service-delivery decisions in a community in order to win approval of their project proposals and maintain an atmosphere favorable to growth. Special districts must balance diffuse constituent preferences for higher impact fees with concentrated lobbying efforts by developers to keep fees low. At the same time, cities and counties become more attentive to demands to redistribute the costs of growth, and governing structure therefore has little influence on policy outcomes. These dynamics are evident only in the fastest-growing communities, but they merit our attention because fast-growing communities are most likely to confront limits in natural resources to support their growing populations.

On the whole, these findings indicate that a special district may be responsive to constituents' preferences on issues outside the district's formal jurisdiction, especially when those outside issues have low salience and the special district's activities receive little attention. What is left unanswered is whether this responsiveness helps or hurts intergovernmental coordination on complex issues. It is impossible to determine from the data whether special districts impose high impact fees in order to cooperate with an overlapping city or county's planning goals or to oppose a progrowth agenda on the part of the government responsible for land-use issues. Informal review of impact fee decisions by water districts and other special districts suggests that the majority of cases can be characterized as cooperative decision making, but conflict is not uncommon. Special districts have substantial influence over development conditions in a community; regardless of zoning and land use ordinances, the value of a potential development site depends on the cost and availability of public facilities. Districts can use their control over these facilities to shape growth policy without the city or county's consent, even though doing so is likely to reduce levels of trust between institutions and raise the transaction costs of coordinating on other policy questions. The analysis suggests that there may be limited cause for concern, however, because districts tend to back off from issues outside their jurisdiction where those issues are higher on the public agenda. Again we see that problem severity plays an important role. On issues that matter the most in a community, governing structure has less effect on policy outcomes.

5
Boundaries and the Incentive to Act Alone

The challenges that growth presents for water utilities reach beyond deciding who pays for pipe extensions to serve new customers. The problem is often one of securing adequate water to meet public demands. Over time, the match between utility service areas and local land use becomes weaker. Economic development, population redistribution across communities, and overall population growth alter levels and patterns of water consumption. Problems also can emerge on the supply side if contamination or overuse reduces the productivity of an existing freshwater source. A community may find its water reliability vulnerable to the growth policies of a neighboring jurisdiction, if the neighbor wants to increase withdrawals from a supply source the two communities share.

In all these examples, problem conditions related to water become more severe over time, as a mismatch develops between a utility's capacity to provide reliable service and the public's consumption demands. An advantage of special district governance is the possibility of creating a good fit between supply and demand at the time of district formation. Designing boundaries for a single function helps reduce negative spillovers into other jurisdictions and takes advantage of scale economies for capital-intensive services such as the provision of drinking water. Nevertheless, the appropriateness of a special district's boundaries may weaken over time as demand patterns shift and new supply constraints emerge. Local water agencies must find ways to extend water service to previously undeveloped areas and to redistribute resources in order to provide water more efficiently. Mismatch between public demands and infrastructure capacity is most evident in rapidly growing regions, where local agencies struggle to keep pace with new development. Depopulation is also costly, however, because it leaves older communities paying to maintain aging systems that are larger than the current population requires (Holst 2007).

Adapting service-delivery patterns to changing territorial demands is a regional problem that requires a regional solution. One strategy for satisfying emerging or heightened demands is to create a new special district. If state law, local interests, or specific problem conditions rule out district formation, then existing governments face the challenge of redistributing their efforts to address the service mismatch. Mechanisms for reaching a solution may include competition or cooperation among governments or some combination of the two. If the challenge is to deliver water to a potentially lucrative area such as an upscale, densely developed new subdivision, governments might compete for the opportunity to annex the community into their service area. Boundary change allows a jurisdiction to act independently to satisfy new patterns of demand. Cooperative interlocal agreements such as partnerships or contracts between governments are a more flexible solution that can address a variety of mismatch problems, including population redistribution and shortages in freshwater supply.

In this chapter, I investigate whether special districts' institutional design affects their choice of strategies for addressing changing problem conditions. Many scholars have reasoned that special districts' boundary flexibility makes them more adaptable to new resource constraints and patterns of demand. Not all districts have equally flexible designs, however. Some districts can easily alter their boundary maps to accommodate community growth; others confront a complex process that requires consent from many local interests. I argue that special districts' response to regional policy challenges is conditional on procedural rules and district resources that influence the costs and benefits obtained from different courses of action. On the whole, districts are more likely to cooperate with their neighbors if they have fewer options for acting alone.

Competition and Cooperation in Addressing Service Mismatch

Even if we assume that special district boundaries have a scale that is appropriate for their function at the moment of district formation—an assumption that ignores a wide range of legal and political factors that might prevent efficient boundary design—discrepancy still can arise over time between a district's actual boundaries and boundaries that would allow it to operate most effectively. Deterioration of a district's water supply is one source of scale problems. Contamination or overuse of a river or aquifer may require a water district to tap a distant supply source, thus increasing costs for pipelines and water storage. More commonly, scale mismatch arises from new demand for water service beyond what existing utilities can satisfy. New demand typically results from residential or commercial development in a growing community, but it also emerges in aging communities where declining productivity of individual wells prompts requests for connection to a community water system.

Changing patterns of water demand can pose a policy challenge whether the demand is located within a public water utility's existing service area or not. Some water systems cannot keep up with rising demand within their boundaries because of insufficient supply or inability to finance infrastructure expansion. In many states, rural water districts are responsible for providing drinking water to the areas surrounding incorporated cities. As development pressures intensify on the city's periphery, rural systems often cannot provide the reliable service that developers and homebuyers demand. Infrastructure limitations including small-diameter pipes, dead-end lines, and aging treatment plants may prevent rural districts from meeting projected consumption levels within their service area even if they have access to adequate supply. In cases where the growth that causes a service mismatch occurs outside any one utility's boundaries, a coordination problem emerges in which local governments must try to ensure water provision without duplicating effort.

Competitive and Cooperative Solutions

The coordinating mechanisms available to address service mismatch can be either competitive or cooperative in nature. One strategy is boundary change, which alters the scale of a jurisdiction to increase efficiency or to address unmet service demand. Annexing new territory into a service area allows a water agency to make productive use of excess capacity and to spread bond debt and system maintenance costs over a broader customer base, producing efficiencies in service delivery and lower rates for existing customers.[1] Consequently, water agencies with slack resources have an incentive to compete with one another over annexable land. Territory that is close to a supply source or part of a potentially lucrative development plan attracts particular interest from water providers. If no utility has clear rights to serve the area, multiple agencies may compete for access. Even territory that falls within a water provider's jurisdiction may be subject to predatory attempts by other utilities. Rural water districts, in particular, routinely face efforts by larger, wealthier urban utilities to "cherry pick" the most lucrative portions of the rural jurisdiction. By peeling off the parts of a rural district's service area that contain compact, high-end development, the urban utility leaves the rural system with responsibility for only the areas that are most costly to serve.

Utilities operated by a city or county usually extend water service into a new territory by expanding their service area outside the jurisdiction of the managing government. Many states permit extralocal service provision by municipal utilities. A city's decision to annex new territory typically would involve more considerations than water service alone. For water districts, in contrast, extending service would more likely entail a boundary change. Changing the district's boundaries means including residents of the new territory as constituents of the district, who then become subject to fees and taxes imposed by the district and able to vote for district officials and bond measures. Municipal annexations often face challenge based on local preferences about the city's racial and socioeconomic balance. Special district boundary change does not affect citizens' definitions of their political community, making it unlikely to encounter any more resistance than would be the case for a simple extension of service (Alesina, Baqir, and Hoxby 2004). Indeed, existing residents of a water district might insist on boundary change in order to ensure that all district customers are subject to the same taxes and fees.

Boundary change helps to solve a service-mismatch dilemma by shaping a single jurisdiction to the scale of the problem or the service demand. A more cooperative strategy is the development of interlocal agreements between two or more water providers. These agreements include contracts for a one-time purchase of water or the establishment of interties between systems that allow water transfers on a continuing basis. They also can allocate responsibility for source-water management, establish terms for the joint construction of distribution or treatment facilities, or formalize the sale of facilities by one water system to another.

By allowing separation of responsibility for water supply and water service, cooperative agreements help solve problems that arise where existing jurisdictional boundaries no longer match the ideal scale for producing or delivering drinking water. Cooperation allows small systems to take advantage of scale economies and large systems to put their slack resources to work. It eliminates service redundancies and may reduce damage from spillovers in a region with fragmented governance. Contracts and interlocal agreements can lower the cost of providing public goods and services within existing jurisdictional boundaries, and they help solve the policy challenges that arise when new demand falls outside the service area of existing water systems.[2] The utility with the most proximate water mains to the area of new demand will usually be the most cost-effective service provider, but it may not have enough water supply for additional customers. A neighboring utility may have slack resources for water production after building in excess capacity in anticipation of future demand.[3] An agreement that brings together the former utility's pipelines with the latter utility's productive capacity offers rewards for both governments and cures the service shortfall. Even boundary review commissions—state and county institutions established to guide local government creation and boundary change—have started to promote intergovernmental agreements as an efficient and less controversial alternative to annexation (ACIR 1992, 33).

The Relationship Between Solutions

Previous studies have characterized special district formation as an alternative to municipal annexation (Carr 2004; Feiock and Carr 2001) or to intergovernmental agreements (LeRoux and Carr 2007) for solving institutional collective action problems.[4] The analysis presented here shifts the focus to special districts' adaptability after they have been established. At the moment of government formation, a district's limited functional scope and the possibility of jurisdictional overlap allow boundary design to fit a public problem. Some analysts have also suggested that the flexibility of special district boundaries over time allows districts to adapt to changing problem conditions and public preferences. Kathryn Foster describes special districts as "geographically adaptable," able to adjust their boundaries in order to accommodate new development or transformation in technology or demand (1997, 97–98). Liesbet Hooghe and Gary Marks (2003) similarly emphasize special districts' flexible design, which allows fluidity in jurisdictional size.

The flexibility of special district boundaries is not universal, however. When crafting legislation that enables formation of new special districts, states define the process for boundary expansion and change. These boundary-change procedures then become part of the package of constraints and opportunities that district officials face when considering strategies for addressing policy challenges related to water service. The formal requirements for boundary change are what Elinor Ostrom (1990, 2005) describes as constitutional choice rules, which affect the selection of mechanisms for overcoming collective action problems. Rules grant voice to or withhold voice from residents of the district and the new territory, and they set procedural requirements for hearings and county review of a boundary-change proposal. Thus, the possibility for a special district to absorb new territory is influenced by a district's boundary flexibility, an institutional characteristic that is defined by state law and that varies across districts both across and within states.

Boundary change addresses service-mismatch problems by creating a permanent commitment for a utility to provide service in a territory. Interlocal agreements promote ad hoc policy solutions that are simultaneously more flexible and less reliable than boundary change. Because cooperative agreements require periodic renegotiation, they may involve higher transaction costs, but they are more likely to produce a solution that is advantageous for multiple governments. They also can help build trust among communities, thereby reaping benefits in the long term. Both strategies offer the potential to produce policy efficiencies and provide more equitable access to water. The question is whether structural rules that make it easier for a district to annex new territory affect its incentives to engage in cooperative policy solutions.[5]

Consider a special district that has slack resources in the form of access to plentiful water supply, excess capacity for water storage and treatment, or a strong financial position that would allow it to expand its infrastructure. When a service shortfall emerges nearby, the district faces a choice between providing services to new customers on its own or entering into a contract to sell that capacity to a neighboring jurisdiction. The effect of boundary flexibility on that decision might operate in either direction. One hypothesis is that boundary change and intergovernmental cooperation are substitute strategies, such that boundary flexibility reduces the likelihood of engaging in interlocal agreements. If the decision between expansion and contracting is a simple trade-off between the costs of building support for annexation and the costs of developing an agreement with another government, then lowering the costs associated with boundary change should produce fewer cooperative agreements. Given the opportunity to act alone, special districts may choose to avoid the costs involved in bargaining with their neighbors. If boundaries are less flexible, however, interlocal cooperation may seem a more attractive strategy.

An alternative hypothesis is that boundary flexibility and intergovernmental cooperation might complement one another by increasing the range of possible solutions to policy dilemmas created by local fragmentation. Boundary flexibility allows governments to position themselves in a way that may promote cooperation—for example, by creating overlap or a common boundary that facilitates labor sharing or infrastructure extension. The possibility of boundary change may create opportunities rather than obstacles for creative intergovernmental problem solving.[6]

Another factor that may influence a special district's cooperative behavior is whether it has resources to meet local demands. Special districts that lack the capacity to provide services to their constituents will be in a weak position relative to their resource-rich counterparts. If districts with slack resources for producing public goods treat boundary change and contracting as substitute mechanisms for distributing those goods, then boundary flexibility will reduce opportunities for resource-poor districts to engage in agreements that allow them to purchase goods and services from their neighbors. If the relationship between competitive and cooperative strategies is complementary, we should see a higher incidence of contracts for both buyers and sellers in a local public economy.

An illustrative case highlights the trade-offs between competitive and cooperative strategies for addressing service mismatch. Over the course of ten years beginning in 1995, the Bexar Metropolitan Water District (BexarMet) near San Antonio, Texas, aggressively expanded its service area. Employing a combination of annexations, lawsuits, and purchase of neighboring water systems, BexarMet spread into three adjacent counties and more than doubled the size of its customer base (Needham 2005b). District officials perceived that expansion would allow them to maintain low rates and pay off debt, but BexarMet's predatory approach to expansion met with hostility from neighboring jurisdictions.[7] A San Antonio representative pursued state legislation that would heighten the hurdles for BexarMet when the agency pursued boundary change, and nearby cities and water districts brought lawsuits charging that BexarMet did not have the legal authority to expand its boundaries in the manner it did.

BexarMet soon discovered that it could not keep up with consumption demands in its swelling jurisdiction. It started rationing water, and the neighboring municipal San Antonio Water System (SAWS) grudgingly allowed BexarMet to tap into its system (Allen 2005; Needham 2005a). When BexarMet attempted further expansion to serve a large, upscale residential community under development in northern Bexar County— all the while increasing its draws from San Antonio's system—SAWS intervened and competed with BexarMet to serve the area, even though SAWS had conceded the territory years earlier (Needham 2005c). The two agencies brought lawsuits against one another; meanwhile, BexarMet was fighting legal battles from its prior annexations. BexarMet's ambitious annexation strategy ultimately weakened the district: two state courts issued rulings prohibiting BexarMet from further expanding its boundaries, and ongoing water shortages forced the district to issue watering restrictions and boiling orders—and even to haul in water by truck—in many of the communities it served (Needham 2006a). In 2006, as state legislators pursued efforts to reorganize the district and rescind the existing board's control, BexarMet sold off five of its territories to SAWS (Needham 2006b).

BexarMet's boundary flexibility allowed it to pursue an aggressive expansion plan, and its competitive stance toward addressing service shortfalls reduced trust in its relations with neighboring water systems and ultimately imposed substantial costs on all water providers in the region.[8] Agencies spent time and money fighting court battles over jurisdiction. Interagency hostilities prevented coordinated planning to address the increasing fragility of the Edwards Aquifer, the region's major supply source. At the time of this writing (2008), BexarMet and SAWS are engaged in a new battle over underground water storage. The agencies had once cooperated in funding studies for a joint storage project, but BexarMet withdrew from negotiations and started pursuing pumping plans that threatened to pull water from the storage project SAWS had developed on its own (Needham 2007b). Meanwhile, BexarMet's customers have endured high and regressive rates, use restrictions, and unreliable water service.

Boundary Rules and Interlocal Agreements

The empirical analysis tests the relationship between boundary flexibility and interlocal cooperation. Analyses of local government organization have highlighted boundary flexibility as an asset of special district governance, but there are reasons to question how flexible special district boundaries are in practice. State enabling legislation prohibits some special districts from overlapping others of the same type, and a recurring criticism of specialized governance relates to the difficulty in terminating a district once the need for its services has declined or disappeared (ACIR 1964; Little Hoover Commission 2000). If special districts are indeed adaptable to changing patterns of demand, it should be possible to eliminate a district once demand has ceased.

Hypotheses

The focus here is on boundary flexibility, or the possibility of boundary change rather than incidence of change itself. Regardless of the frequency with which special districts adjust their territorial reach, it is the obstacles they face in doing so that define the flexibility of their boundaries. State law makes it more or less difficult for a district to pursue annexation as a strategy for addressing supply shortages or unmet service demands. A district that has fewer obstacles to boundary change will perceive annexation as a more viable solution, whether the district actually attempts larger or more frequent annexations or not.[9]

Boundary flexibility therefore refers to the stringency of state rules guiding the process of special district boundary change. If a relationship exists between boundary flexibility and intergovernmental cooperation, district officials will consider the relative ease of both strategies when deciding how to solve a service-mismatch problem. Restrictive rules will make boundary change a less attractive option for district officials. If cooperation is a complementary strategy, restrictive rules also should suppress interlocal agreements. If the strategies are substitutes, agreements should replace boundary changes where rules are stringent.

With regard to the BexarMet case, a substitute relationship between boundary flexibility and interlocal agreements would suggest that BexarMet's freedom to annex new territory and expand its service area stifled development of a cooperative approach to drinking water provision in the San Antonio region. Had BexarMet faced higher costs in changing its boundaries and eventually overextending its resources, agencies in the region may have developed a joint solution to the service-mismatch problem, with potential benefits for economic efficiency, aquifer protection, and service reliability. Not all interlocal agreements reflect cooperative relationships among their signatories, however. Recall that BexarMet did not have a sufficient water supply to keep up with its rapid expansion. San Antonio reluctantly allowed BexarMet to tap into its water system as an emergency measure, and BexarMet's reliance on SAWS water persisted for years. The case therefore may provide evidence of a complementary relationship between flexible boundaries and intergovernmental agreements. Boundary flexibility might promote interlocal action by expanding the universe of potential policy solutions, or—as in the San Antonio case—it may force interlocal reliance by allowing utilities to engage in poor planning.

Model
The analysis estimates the effect of water district boundary flexibility on the incidence of intergovernmental agreements. Included in the analysis are independent special districts that report water supply as their primary function. Data on intergovernmental agreements come from the 2002 *Census of Governments* (U.S. Census Bureau 2005a). The models estimate effects for two different dependent variables: local intergovernmental revenues and local intergovernmental expenditures. Both are coded as dichotomous variables; an intergovernmental agreement exists when a water district reports any local intergovernmental revenue or spending. The dependent variable measures only intergovernmental agreements related to water functions. Across all functions, special districts are less likely than their general-purpose counterparts to establish intergovernmental agreements: 32 percent of special districts report some local intergovernmental spending or revenue, compared to 54 percent of cities and towns. As Robert Stein (1990) shows, the nature of a good can influence the likelihood of alternative service provision. Interlocal cooperation is rarer for water than for other functions, and only 12 percent of water districts participate in interlocal agreements related to water.[10]

The source for data on boundary change rules is state enabling legislation. State statutes that enable special districts contain procedural rules for changing district boundaries subsequent to district establishment. The procedural rules parallel requirements that the ACIR has identified and coded for municipal annexation: majority approval of the boundary change by residents of the district, majority approval by residents and/or landowners of the territory to be added to the district, organization of a public hearing, and approval from a county governing authority (ACIR 1992).[11] These requirements are combined into an index measuring the stringency of rules for special district boundary change.[12] In addition, I measured whether a requirement exists for new territory to be contiguous with a special district's existing jurisdiction.

Careful examination of procedural requirements for water district boundary change reveals that districts are not as adaptable as some analysts have suggested. Once a water district has been established, district officials and other local actors may face important hurdles to expanding

Table 5.1
State Boundary-Change Rules

	Municipal annexation index (ACIR)	Water district boundary-change index	Water district contiguity requirement
Alabama	1	1	1
California	1	4	1
Colorado	3	2	0
Idaho	0	2	0
Kansas	1	3	1
Kentucky	3	2	0
Missouri	0	2	1
Montana	3	3	0
Nebraska	0	2	0
New Hampshire	0	3	0
Oregon	2	3	0
Tennessee	3	2	0
Texas	2	3	0
Utah	1	2	0
Vermont	0	2	1
Washington	3	3	1
West Virginia	2	2	0

the district's jurisdiction. Table 5.1 shows boundary change rules for the seventeen states included in this analysis.[13] The first column indicates the state's score on the four-point index for municipal annexation rules, using data collected by the ACIR. The second column includes the parallel score for water district boundary-change rules. Because these rules may vary across district types within a state, table 5.1 shows the average boundary-change score for water districts in the state, with more common district types receiving more weight in the state-level score. The number of procedural requirements ranges from one in Alabama to all four in California, demonstrating the variation in special district boundary flexibility. Moreover, in the majority of states, procedures for water district boundary change are more restrictive than procedures for municipal annexation. Expectations for contiguity are relatively rare among special districts, however. These rules appear in the table's final column.

The majority of water districts in eleven of the seventeen states may add territory that is not adjacent to the current district.

Because boundary flexibility is measured by the number of rules restricting boundary change, a more flexible special district is one with a lower score on the boundary rules index. If boundary flexibility is a substitute for intergovernmental cooperation, then we should expect a positive relationship between boundary rules and incidence of a revenue-generating interlocal agreement. A negative effect would suggest a complementary relationship in which restrictions on boundary change reduce agreements by suppressing opportunities for creative interlocal collaboration. A complementary relationship also would produce a negative effect for boundary rules on the existence of expenditure agreements. If expansion and cooperation are substitutes, then higher hurdles to expansion should increase local governments' ability to engage in expenditure contracts.

Other variables in the models control for fiscal, intergovernmental, institutional, and problem-severity conditions that also might affect the likelihood that water districts will engage in interlocal cooperation.[14] Fiscal variables measure a special district's capacity both to fund its own water supply functions and to offer resources to its neighbors as well as the tax burden on area residents. System size—measured in terms of district expenditures—should affect contracting if small districts are less self-reliant than their larger counterparts. Two variables measure debt: a measure showing whether the water district reports debt financing as part of its operations and a measure of the district's outstanding long-term debt. The former indicates whether the district perceives the construction of capital facilities as one of its primary functions, suggesting that it may have extra capacity to share with neighboring jurisdictions. The latter measures overall indebtedness, which should promote contracting because of the limits it puts on a district's ability to finance new construction. In addition, the models account for property taxes per capita imposed by the state and all local governments located within the district's home county in order to account for vertical competition for tax revenue. A heavy tax burden on the local population will make boundary change less attractive to special districts, which may make districts more likely to cooperate rather than build and operate their own expensive facilities.

Because interlocal cooperation is more likely to occur where local governments have access to a larger number of potential partners, the models include a measure of the number of local governments located within the special district's home county. Multicounty districts should be more likely to find opportunities for collaboration, and districts with boundaries that correspond to a single city or county might enjoy a long-term cooperative relationship with the overlapping general-purpose government. Proportion spending on water indicates the proportion of the district's total current general expenditures dedicated to water, because the degree to which a district specializes might influence its opportunities for collaboration. Finally, two variables in the models control for the use of other possible coordinating mechanisms in the face of interlocal policy challenges: a four-point index measuring the stringency of procedural requirements for municipal annexation and the change in the number of special districts located in the county between 1992 and 2002 per 10,000 county residents.

The analysis also accounts for the proportion of the special district's governing board that is elected rather than appointed to office. This variable might operate in either of two directions. Elected boards should have a stronger incentive to seek out contracts in order to achieve policy efficiencies that lower the tax and fee burden on district constituents. However, appointed boards might have a broader geographic scope because board members are accountable to elected officials who may represent a large jurisdiction. Moreover, appointed boards may include representatives from other local governments that might be potential partners. The final set of control variables addresses the seriousness of local water-policy problems. By heightening the potential benefits of cooperation, problem severity may allow the emergence of cooperative action among governments. Indicators of problem severity include mean annual precipitation and mean maximum daily temperature, as well as percentage change in population of the water district's home county between 1990 and 2000.[15]

Results

Results from the analysis reveal the importance of both structural rules and problem context in influencing special districts' cooperative behavior. Boundary flexibility affects a district's likelihood of engaging in interlocal agreements, but the nature of that relationship depends on whether a special district is offering or seeking services. Stringency of boundary-change rules has a positive effect on interlocal contracts when the contracts generate revenue; for expenditure contracts, boundary rules have

Table 5.2
Establishment of Interlocal Agreements on Water

	Revenue agreements	Expenditure agreements
Boundary-change rule index	.022***	−.020*
Contiguity requirement	−.037***	−.016
Fiscal variables:		
Current expenditures (log)	.011	−.007
Debt finance	.014**	.007
Long-term debt (log)	.038*	.015*
Property taxes per capita	.003	.004***
Intergovernmental variables:		
County local governments	−.001	.000
Multicounty special district	.008	.015*
Common boundaries	.034**	−.013
Proportion spending on water	.028**	.025***
City annexation rules	−.013	.017
District formations	−.004**	.000
Institutional variable:		
Proportion elected	−.024	.027***
Problem severity variables:		
Precipitation	.004	−.000
Temperature	−.006	−.011**
Population growth	−.007	.002

Notes: Cell entries show the difference in predicted probability of interlocal cooperation associated with a shift from the twenty-fifth to the seventy-fifth percentile value of each independent variable (or from 0 to 1 for dichotomous variables and proportion of spending on water), fixing all other variables at their mean values. Probabilities are based on estimates from a complementary log-log model with observations clustered by special district type within a state. The model appears in the appendix. Estimates are significant at $^*p < .10$, $^{**}p < .05$, $^{***}p < .01$ (two-tailed).

the opposite effect. Table 5.2 presents the results as marginal effects, and figures 5.1 and 5.2 depict the probability that cooperation will emerge across the range of boundary restrictiveness. Increasing the number of procedural hurdles for district boundary change from the twenty-fifth to the seventy-fifth percentile—from two to three rules on the boundary index—boosts the likelihood that a district will participate in a revenue-generating contract by 2.2 percentage points, holding all other variables at their mean. Moving from the minimum of a single rule to the

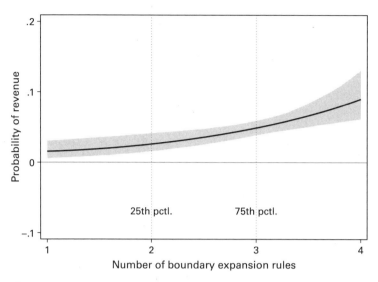

Figure 5.1
The relationship between boundary rules and revenue contracts. *Source:* Complementary log-log estimates in the first column of table A3.1. Gray band indicates 95 percent confidence interval.

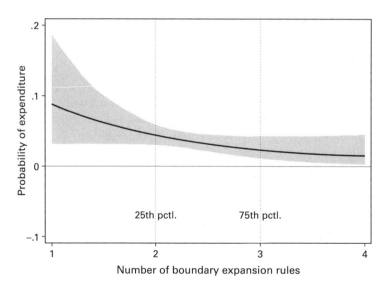

Figure 5.2
The relationship between boundary rules and expenditure contracts. *Source:* Complementary log-log estimates in the second column of table A3.1. Gray band indicates 95 percent confidence interval.

maximum four rules produces a 7.5-point increase in the probability of cooperation. These results are significant with a high degree of confidence, and the effect sizes are substantial considering that just 6 percent of sampled water districts participated in revenue contracts.

This finding supports the hypothesis that boundary flexibility and intergovernmental cooperation are substitute strategies for addressing local coordination problems. When a water district has surplus supply or other resources, it can put those resources to use by expanding its own boundaries to enlarge its customer base, or it can establish a cooperative agreement that allows other local governments to take advantage of its excess capacity. Depending on the nature of the specific policy problem, either of these strategies might be the more efficient and durable policy solution. But the analysis suggests that the two are in fact alternative solutions—where boundary change is a more feasible strategy, the incidence of interlocal cooperation declines.

A requirement that additional territory be contiguous to a water district's existing jurisdiction has the opposite effect: it reduces the probability that a district will enter into a revenue agreement. All other boundary rules held constant, special districts that can expand into areas noncontiguous to their existing boundaries are more likely to enter into revenue-generating water contracts. This result may signify diversity in the types of coordination challenges facing local governments. It is possible that special districts view annexation and contracting as substitutes in the typical case of service extension beyond district boundaries, but when faced with more complex and far-reaching collective action problems, creative boundary manipulation helps promote cooperative policy solutions. Another possible explanation for this finding is that special districts will more likely require assistance from their neighbors when they overextend themselves by serving noncontiguous territories. The latter explanation would help account for outcomes observed in the BexarMet case.

In contrast, boundary rules have a negative relationship with special district spending on interlocal cooperation. Restricting the flexibility of a district's boundaries reduces the probability that the district will enter into a contract to purchase water supply or services from another local government. This result suggests that districts with stable boundaries are less likely to confront service-mismatch problems that require them to seek assistance from neighbors. Without the opportunity to grow through territorial expansion, districts may be better able to develop their infrastructure on pace with changes in demand. However, this finding is weaker than the finding for revenue agreements. The estimated size of the effect is nearly equal to the positive impact of boundary-change rules on revenue-generating cooperation, but the effect is only weakly significant, and it is not robust to changes in the estimation strategy. Contiguity requirements have no apparent effect on establishment of expenditure contracts.

Among fiscal considerations, debt affects the incidence of both types of cooperation. Districts that report debt financing as one of their primary functions are more likely to engage in revenue agreements, and higher levels of outstanding debt promote contracting from both sides. The size of district budgets has no apparent impact. The latter result counters expectations that districts with more resource capacity would provide services to other localities, but the coefficient for outstanding debt probably captures this effect of system size.[16] The county's property tax burden makes it more likely that a water district will contract for supply or infrastructure rather than develop new capacity itself, but it does not affect the incentives for revenue contracting.

Intergovernmental variables have somewhat more influence on revenue than on expenditure agreements. Having boundaries that correspond to a city or county increases the probability of providing supply or services to another local government by a substantial three and a half percentage points. Recent special district formations in the county have a small but significant negative effect on the likelihood of revenue contracting, supporting the notion that special district formation is another substitute to contracting for the provision of local services. More dedicated focus on water boosts contracting of both types, but the singular focus of most water districts limits this variable's utility in explaining overall patterns of cooperation. Although a hypothetical shift from 0 to 1 in proportion spending on water functions would produce approximately a two and a half point increase in the probability of engaging in cooperation, in fact more than 75 percent of sampled districts dedicate all their spending to water. For districts pursuing expenditure contracts, a multicounty jurisdiction exhibits a weakly significant positive effect on formation of agreements, perhaps owing to the availability of more potential partners.

Election of special district officials has a sizeable and significant positive effect on cooperation for districts requiring services, suggesting that politicians may perceive contracting as an effective strategy for addressing service gaps and maintaining public support. Surprisingly, conditions

Table 5.3
Effects of Specific Boundary-Change Rules on Interlocal Agreements

	Revenue agreements	Expenditure agreements
Boundary-change rules:		
Majority approval within district	.042**	−.030**
Majority approval in new area	.007	.003
Public hearing	—	—
County approval	.027	−.021*
Contiguity requirement	−.047*	−.015

Notes: Cell entries show the difference in predicted probability of interlocal cooperation associated with each individual boundary rule, fixing all other variables at their mean values. Control variables are omitted from the table. Probabilities are based on estimates from a complementary log-log model with observations clustered by special district type within a state. Estimates are significant at *p < .10, **p < .05, ***p < .01 (two-tailed).

that affect the severity of water supply issues in a community do not have a clear and consistent relationship with cooperative behavior. None of the problem-severity variables has a significant impact on the probability that water districts will contract out their own services through revenue-generating agreements. A hot climate has a modest effect on the likelihood of contracting to obtain services, but the estimate operates in the opposite direction as predicted. Water districts in hotter regions appear to be less likely to contract with their neighbors in order to cope with mismatch between water supply and patterns of demand.

Table 5.3 unpacks the boundary-change index to show the effect of specific boundary rules.[17] Results indicate that the referendum requirement within a district's existing service area is the biggest hurdle to boundary change and therefore the most important influence on adoption of revenue-generating interlocal agreements. With other boundary rules and control variables held constant, requiring majority approval within the district for a boundary change increases by 4.2 percentage points the likelihood of contracting to provide goods or services. It is the same rule that reduces the likelihood of a district's reporting interlocal spending, although the requirement for county approval exerts an additional weak negative effect on expenditure agreements.

These results suggest that a different process is at work for special districts with and for those without slack resources for the production and provision of drinking water. Districts with extra capacity face a choice between acting alone and acting jointly when deciding how to distribute their resources. Raising the cost of annexation makes interlocal cooperation a more attractive strategy for addressing service mismatch and putting slack resources to use. At the same time, boundary hurdles may reduce the probability that a resource-poor district will contract with its neighbors to obtain goods or services. Inflexible boundaries may help prevent cases like the one in San Antonio where a water district's resources cannot keep up with its rapid territorial growth. In those cases, strict boundary-change rules help protect service quality for constituents and avoid the development of new mismatch problems. However, a narrower range of mismatch solutions exists where boundaries are inflexible. Districts lacking slack resources might not be able to contract for assistance from their neighbors if they cannot adjust their boundaries to create territorial overlap. The net result of boundary flexibility is mixed. Flexible boundaries make it easier for districts with resources to forgo cooperation and act alone, but they also may allow for creative and cooperative solutions where acting alone is not an option.

Conclusion

Special district governance is not a cure for regional policy problems, nor can it be implicated in consistently worsening those problems. In some cases, special districts help avoid or mitigate policy challenges related to scale or externalities. In other cases, they contribute to fragmentation of local authority and problems of collective action. The important question is how special districts confront regional problems that arise. The analysis presented in this chapter shows that institutional design and district resources have an important influence on how districts choose to address policy challenges that cross jurisdictional lines.

Interlocal agreements are a flexible policy tool that can help localities close service gaps, overcome mismatch in the distribution of resources and resource demand, and improve efficiency in service delivery. Yet the transaction costs for developing an agreement may be high, and the possibility of renegotiation or exit increases uncertainty for participants and hinders long-term planning. When public officials consider entering into a cooperative agreement, they balance its expected benefits against the costs of developing and maintaining the partnership. My findings indicate that officials also consider the costs of other strategies that might allow them to achieve the same policy goals. Boundary flexibility is part of a special district's institutional design and determines the ease with which a district can act unilaterally to solve a regional problem. The costs of acting alone affect the likelihood of cooperation among neighbors.

Cooperation is conditional not only on the procedural hurdles a district faces in changing its own boundaries, but also on the type of policy challenge a special district confronts. Among districts with excess capacity for production and provision of drinking water, boundary change and interlocal contracting are substitute strategies. The more difficult it is to annex new territory and secure a permanent market for the district's resources, the more likely it is that the district will establish interlocal contracts allowing neighboring governments to utilize those resources. For districts seeking resources from an interlocal agreement, boundary change and contracting are complementary. Flexible boundaries appear to offer more opportunity for districts to identify partners and establish agreements.

Although boundary change and interlocal contracting appear to be substitute strategies for water districts with excess capacity, it is still possible that districts pursuing boundary change have cooperated informally with their neighbors. Faced with a water-service mismatch, localities in a region may reach agreement that expanding the boundaries of an existing water district is the best mechanism for internalizing spillovers or addressing service gaps. Acting alone does not necessarily mean acting without consultation. The data analyzed here do not allow inferences about the process of policy choice in these scenarios of fragmented authority. Case studies presented in chapter 6 address the dynamics of interlocal competition and cooperation in more detail.

6
Fighting over Land and Water: Venues in Local Growth Disputes

The Bexar Metropolitan Water District outside San Antonio made a series of management mistakes beginning in the 1980s that produced the service gaps described in the previous chapter. After aggressively expanding its territory through the purchase of small water systems, the district discovered that it lacked the water supply and infrastructure needed to satisfy customer demands. It began buying up water rights in the region and undertook ambitious infrastructure projects to allow withdrawals from surface sources to supplement its water from the Edwards Aquifer. These efforts fell short, however, and water pressure and quality continued to decline in parts of the district's service area. BexarMet also fell deeply into debt, forcing the district to keep water rates high and prohibiting further infrastructure improvements. By 2006, BexarMet was enforcing usage restrictions and issuing boil orders while it fought off a takeover attempt by the Texas state legislature.

Underlying BexarMet's troubles were the challenges inherent to intense growth in a region with limited water resources. The population of the San Antonio metropolitan area grew by nearly 50 percent between 1980 and 2000. The growth rate was slightly higher outside the city, in areas served by BexarMet, than in the service territory of San Antonio's municipal water utility. In the three surrounding rural counties where BexarMet's service area extended, the combined twenty-year growth rate was 86 percent. This growth has been a strain on the Edwards Aquifer, the primary source for the region's drinking water supplies. Heavy pumping during a drought in the 1980s caused the aquifer level to drop, reducing flow at the major springs and jeopardizing endangered species. In 1993, under pressure from a federal court ruling, the Texas legislature established the Edwards Aquifer Authority to manage and protect this groundwater resource. The authority has capped the amount of water that may be withdrawn from the aquifer, and it has a mandate to set further limits as needed to maintain minimum spring flows. Should a severe drought occur, studies predict that limits may be set as low as half the withdrawals allowed in 2007 (Texas Water Development Board 2006). The state water agency estimates that failure to make alternative plans for potential drought may cost the San Antonio region more than ten thousand jobs and $665 million in lost income by 2010, with greater economic impacts further into the future (Texas Water Development Board 2006).

BexarMet officials believed they had a responsibility to support the San Antonio region's rapid growth ("Adding Water" 2004; Aldridge 2000). The district could certainly have done a better job in managing its system and ensuring a safe and reliable water supply for its customers. San Antonio's municipal-owned utility accommodated similar growth rates and managed to hold down water prices, maintain high service quality, and initiate long-range planning and conservation programs.[1] Nonetheless, BexarMet's poor decisions came in response to severe growth challenges that confront water systems throughout the nation. As utilities begin to face permanent limits on their water supplies, the decision to extend service to new households and businesses carries the potential cost of reducing service reliability to all customers during periods of drought. Even in areas with abundant freshwater resources, securing water for new development may require bargaining over rights and infrastructure to transport the water from its origin to the point of use.

What happens when water is not available to serve planned residential growth? What are the possible policy solutions? Who gets to decide whether development proposals get approved? And does the organization of local government affect how water issues get considered in land-use planning decisions?

This chapter explores the effect of specialization on the politics and outcomes of disputes over land use and water. In communities with consolidated governance, a single city or county oversees both functions. Incorporating water as a factor in land-use decisions requires increased communication and cooperation among municipal departments, but the same public officials who approve new development will ultimately be held accountable if this decision has negative impacts on water supply or quality for existing residents. Where a general-purpose government makes land-use decisions and a special district provides water, these separate, independent governments have no obligation to cooperate. Local officials who approve new development can avoid responsibility for its effects on water resources, and water district officials are subject to decisions made by politicians in cities and counties about whether to add to demands on the water system. When water issues enter into disputes over growth, how do these governing arrangements affect the politics and outcome of the dispute?

I focus on a series of local disputes in California and Pennsylvania to examine how governing structure influences local governments' incentives to cooperate on complex policy questions involving water and land use. Drawing on interviews, lawsuit briefs, and other qualitative data, I find that the multiplication of policy venues adds considerable costs to the resolution of such questions, but the costs do not systematically favor either side in a growth dispute. Specialized governance shifts negotiation over water and land use into the public sphere, creating opportunities for participation by the public, neighboring governments, and interest groups. These actors often take advantage of crosscutting jurisdictional boundaries that allow disputes to be transferred between decision-making venues. Participation by outside groups makes it more difficult for land-use and water authorities to negotiate and compromise, which may reduce trust and policy coordination in the long term. On the whole, I find that specialized governance can pose a significant hurdle to cooperative policymaking on policy questions that cross functional boundaries. Local policymakers' ability to overcome this hurdle depends in part on the local salience of the issues under consideration and on the institutional factors that affect the mobilization of local interests.

Integrating Water and Land-Use Planning

Water and land-use planning traditionally have proceeded along separate paths. Cities and counties approve a development project and expect the local water agency to extend service to new residents. Facing additional demands on their water systems, utilities with limited supply must seek out new sources to avoid jeopardizing service to existing businesses and residents. Public-utility law in many states explicitly defines a public water system's duty to serve all customers within its service area, making it difficult for utilities to control new demands on their systems (U.S. EPA 2006).

A water provider occasionally resists by refusing to allow new users to tie into its system. A notable and particularly long-lasting water-connection moratorium has been in place for almost forty years in the unincorporated community of Bolinas in coastal Marin County, California. In 1971, the Bolinas Community Public Utility District declared a water-shortage emergency and prohibited further connections to the system. The moratorium has allowed the community to maintain a rural, anachronistic 1960s character while sitting on some of the West Coast's most desirable real estate. After spending eleven years and nearly $2 million defending the moratorium against a legal challenge, Bolinas has no plans to seek out new supply sources that would allow the ban to be lifted. For the near future at least, development in Bolinas will be limited by the 580 water meters that may be connected to the community's system. One of these meters sold for $310,000 in a 2005 auction (Bernstein 2005).

Most moratoria are not so long lasting. Utilities typically declare only temporary halts on connections until new supply or infrastructure comes on line or drought emergency conditions pass. Water providers in rapidly growing regions will sometimes enact a moratorium to allow time for long-range planning and studies of system capacity. Because such moratoria usually last only a short time, it is difficult to know how widespread their use might be. Thirteen percent of California cities and counties reported in 2004 that at some point they had implemented building moratoria in response to water supply concerns, most commonly during droughts that affected the state in the late 1970s and late 1980s (Hanak 2005). Moratoria are less common but still seen in regions where water is more plentiful.[2] As it becomes more difficult for utilities to obtain new water supplies, temporary moratoria may be enacted more frequently.

Although most building and connection moratoria are sincere responses to conditions in a community's water supply, the power of moratoria is apparent to local actors who oppose growth. Approval for new residential development is useless without assurance of potable drinking water. Developers may be willing to provide roads, schools, or other infrastructure for new houses, but rarely will they build without securing access to a public water supply (Fulton, Pendall, Nguyen, et al. 2001). For those politicians and residents who seek to block growth, stopping the release of new connection permits might have the most legal and procedural force among the set of possible tools for achieving their goal (Frieden 1983; Herman 1992).

Short of a complete moratorium on new connections, other options for reducing demands on a water system—or for stalling new development in a community—include annual caps on new connections and refusal to expand a utility's service area. Alternatively, a utility attempting to accommodate growth might require developers to identify and pay for a supplementary water supply, which the utility then treats and pipes to the new homes. This approach allows the utility to avoid the search and purchase costs for finding surplus water by shifting those costs to developers and prospective homebuyers instead. It puts developers in the role of water prospectors, scouting for an increasingly scarce natural resource (Kasler 2002).

An ad hoc system of connection restrictions and water transfers jeopardizes local freshwater supplies and creates risks for landowners, developers, home buyers, and existing residents. When land-use planning proceeds without consideration of water resources, utilities might continue tapping declining aquifers and overstrained streams. Development plans often do not provide for adequate groundwater recharge or watershed protection. Transfers of water from a distant source require investments in pipelines and new water storage or payments to third parties for water conveyance in an arrangement known as *wheeling*. Transfers also can have significant environmental and economic impacts in the region of origin. If water supplies still fall short of local demand, new homebuyers may find they have invested in dry lots, and long-time residents may confront usage restrictions and deteriorated water quality. Where utilities attempt to avoid these outcomes by refusing to add new connections, they often face legal challenges from developers and landowners seeking to protect their investments.

Another problem with individually negotiated solutions is that they can be a product of local politics as much as of water scarcity. Because connections moratoria are an attractive growth-management tool, there is no guarantee that the communities limiting new connections face real constraints in their system capacity. Water might be an excuse to keep out unwanted residents from communities that in fact have the resources to support a larger population. In contrast, as the BexarMet case reveals, some utilities facing resource constraints will seek out growth, believing that a larger customer base will allow them to capture efficiencies and lower water costs to existing residents.

The disconnect between water and land-use planning poses an obstacle to the sustainable provision of drinking water in communities where people want to live. In response, many western states have enacted laws requiring that local development approval be conditional on demonstrated adequacy of water supplies. Although the specific elements of state water-adequacy laws vary, most dictate that local water utilities—sometimes in cooperation with a state agency—must assure long-term water availability before a land-use authority can approve a large-scale subdivision proposal. The regulations force some level of consultation between water and land-use officials, even if they do not impose full integration of planning processes between the two.

Water-adequacy laws have the potential to benefit developers and landowners by providing an early signal that water might not be available for a project, yet the development community nevertheless opposes these initiatives based on concern about overzealous interpretation by local actors pursuing an antigrowth agenda. Water-adequacy reviews can limit growth even where water supplies are plentiful, because no objective standards exist to define "adequacy" (Hanak and Browne 2006). Uncertainty over future groundwater yields allows for differing interpretations about whether supplies are sufficient to meet projected demands. Judgments about adequacy also hinge on the drought level a community wants to prepare for and expectations for conservation by existing residents. Almost any community can declare water supplies inadequate by relying on conservative estimates of groundwater yields, planning for worst-case drought scenarios, and requiring little sacrifice from existing residents should those scenarios come to pass. Of course, the same holds true in reverse: water agencies may be even more likely to overestimate their capacity to accommodate new growth.

In general, however, water-adequacy screening aims to promote coordination between water and land-use planning processes in an attempt to reduce water shortfalls that can hurt both developers and residents. Early evidence after enactment of California's laws indicates that state screening requirements can play an important role in triggering consideration of the supply impacts of large development projects in areas where local review processes are weak (Hanak 2005). These processes are particularly weak where authority for water and land use is formally divided between governments. Water and land-use officials might have different preferences regarding community growth. Even if they share the same goals, the costs of exchanging information and reaching agreement on specific development proposals are higher where lines of authority reside in separate governments.

Results from a 2004 survey of California land-use planners demonstrate the coordination problems that arise from specialized water governance (Hanak 2005; Hanak and Simeti 2004). The results are shown in

Fighting over Land and Water: Venues in Local Growth Disputes 129

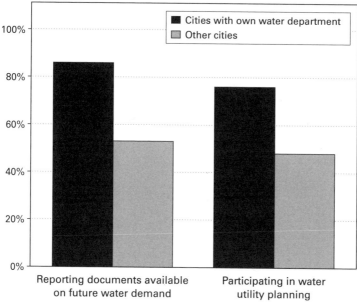

Figure 6.1
Integration of water and land-use planning across forms of water governance.
Sources: Hanak 2005; Hanak and Simeti 2004.

figure 6.1. Among planners in cities with their own municipal water departments, 86 percent reported that they were aware of water-planning documents projecting the impact of demographic change on water demand. Only 53 percent of planners in cities without municipal water were aware of such documents, even though the researchers found that water districts were more likely than municipal utilities to prepare these documents. Specialized governance of water affects not just information flows, but also the incidence of joint planning activities. Three-quarters of planning departments in cities with water utilities reported participating in water-planning activities, compared to less than half of planning departments in cities with outside water providers. With low rates of coordinated planning in communities that have specialized water governance, state water-screening requirements may be the best option for ensuring that water issues get addressed before development proceeds.

Given the potential impacts of land-use decisions on water supply and the ongoing discord about growth in many communities, it is not surprising that disputes sometimes arise over issues of land use and water. Local actors have a number of incentives for introducing water issues into a community debate over growth. The goal may be to reduce a proposed project's impact on water resources, shift the cost of conservation or system extensions to incoming residents, or block a proposed development entirely. Interests supporting growth must overcome challenges based on water issues, and local officials must consider how heavily to weigh water-based arguments in deciding whether to approve new development.

Conflicts involving land use and water are complicated even where lines of authority are clear. They evoke competing visions of community identity and disagreement about how to manage scientific uncertainty and local fiscal limitations. Participants enter these disputes with conflicting goals and different ideas about the meaning and value of water itself. Developers seek to harness and direct water so it can meet the demands of homeowners and promote economic prosperity. Environmentalists favor maintaining natural water flows in order to protect species habitat and the scenic and recreational value of a water source. Residents who seek to maintain a community's existing character might see water availability as the best opportunity to block development they oppose based on broader considerations. Other residents may view water as a human right and oppose any effort to limit access or usage. Water-agency officials tend to espouse a conservative set of values, aiming above all to deliver water reliably and to attract little public attention (Lach, Ingram, and Rayner 2005). As the cases discussed next reveal, divided governance over land use and water can aggravate rather than ameliorate the conflict among these competing values and goals, and it may spark a battle for control over decision-making authority.

Case Background

To evaluate the influence of arrangements governing water on disputes over water and land use, I investigate a series of disputes that occurred between 1991 and 2004 in four regions, two in northern California and two in Pennsylvania. The bulk of the analysis focuses on the California cases, but I include the Pennsylvania cases to highlight the dynamics of policy disputes involving land use and water in a region where water historically has had little salience. In all of these disputes, a general-purpose local government faced a decision about growth in the context of uncertain water availability. In one region, both the retail and wholesale water providers are part of multifunction governments; in the three other regions, special districts oversee water provision. For the disputes that took place where responsibility for water and land use is divided between governments, I also examine decisions made by water districts about whether to extend service to the proposed development.

The qualitative data examined in this chapter come from a variety of sources. I make extensive use of newspaper accounts to establish the sequence and timing of events and to gain insight on participants' positions and motivations at the time. I also rely on primary data sources including lawsuit briefs, environmental documents, memos and meeting minutes from public agencies, and interest-group literature. In addition, much of the data for the California cases come from semistructured, open-ended interviews with city council members, county supervisors, water district officials and staff, interest-group representatives, and residents involved in the conflicts over land use and water.[3]

California

Authority over land-use policy in California, as in the rest of the nation, has historically resided firmly with municipalities. Counties take the lead in planning for unincorporated areas. The planning process laid out in California law provides ample opportunity for citizen participation and comment. Every city and county must have a comprehensive, long-term general plan to guide future development, supplemented with more detailed specific plans. At every decision-making stage, local governments must circulate planning documents, hold public hearings, and otherwise create opportunities for residents to express their opinions. Open meeting and sunshine laws shed further light on the proceedings.

In addition to these requirements in planning law, the environmental review process outlined by the California Environmental Quality Act provides a means for citizens to challenge land-use decisions through the administrative sector and in court. The act requires public agencies to prepare an environmental impact report (EIR) and to create opportunities for public review and comment on projects that are expected to have significant environmental impact. In California, EIRs are required for all but the smallest projects; the state's environmental quality law applies more broadly than its federal counterpart. The process invites the public to comment not only on the merit of the project itself, but also on the completeness and adequacy of the project review. Environmental reviews are subject to challenge in court, and these legal challenges have become the primary weapon in local battles over growth.

The California disputes presented here took place in a context where growth was a longstanding, heavily debated public issue, and concern about water supply would periodically appear on the public agenda. Local government decision making occurred through fairly open processes, and residents were accustomed to expressing their preferences about growth and development by voting on direct ballot measures. These conditions clearly do not apply universally to all disputes over land use and water. In many communities, decision-making processes are more closed, and growth issues are not a matter of public debate. But with the spread of direct democracy and sunshine laws and with growing attention to the consequences of sprawl nationwide, these conditions are becoming more common. In California, issues of water availability play an increasingly important role in decision making about proposed development—in whether the development goes forward and in what its scope or its cost will be. I assess how specialized governance contributes to shaping these outcomes.

Dougherty and Tassajara Valleys, Contra Costa County The locations of the California case studies appear in figure 6.2. The first region is an unincorporated area in southern Contra Costa County, on the eastern periphery of the San Francisco Bay Area. The disputes in this area revolved around three proposals to build large housing developments on previously undeveloped agricultural lands in Dougherty and Tassajara valleys. In a region with specialized governance of retail drinking water service, both valleys fell outside the designated ultimate service boundary of any water utility. During the period that the three projects were under consideration, the largest and most proximate water district, the East Bay Municipal Water District (EBMUD), had a standing policy not to expand its ultimate service boundary, arguing that it lacked sufficient water supply to serve the proposed developments. Developers backing the projects needed to identify a water source and a willing provider. They also needed to overcome local activists and neighboring communities' efforts to block valley development.

After decades of considerable population increase, growth was a highly salient policy issue in Contra Costa County. Dougherty and Tassajara valleys are located on the periphery of the rapidly expanding Tri-Valley region that has added more than one hundred thousand new residents

Fighting over Land and Water: Venues in Local Growth Disputes   133

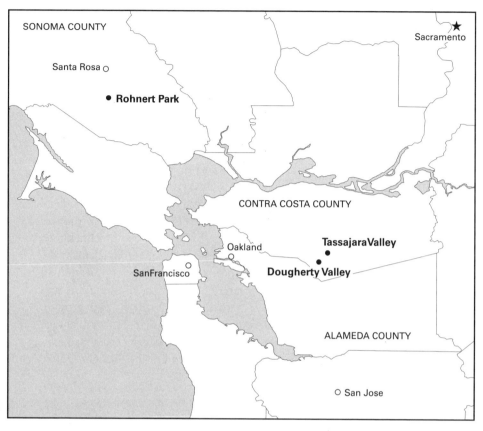

Figure 6.2
California case study locations.

each decade since the early 1940s. Environmentalists had consistently opposed development in this region, and over time they were joined by residents who were beginning to experience traffic congestion and other effects of explosive growth. Residents did not seek to block growth altogether, however. In 1990, facing competing growth-control measures on the ballot, Contra Costa County voters approved the plan endorsed by the county board of supervisors and rejected a more restrictive measure that would have contained future development within existing city limits. The winning proposal allowed development in unincorporated areas but established an urban limit line to direct growth to the areas around existing cities. It also required the county board to get assurances for infrastructure and local services, including water service, before approving new development.

The severity and salience of water issues varied over the period of this study. A statewide drought from 1987 through 1991 raised public awareness about water scarcity and volatility in the state's supply, and it renewed concerns about the adequacy of EBMUD's water supply to withstand extended dry periods. A decade earlier, another drought had reduced the district's reservoirs to just 2 percent of their capacity. An EBMUD staff member recalls, "That experience, I think, more than anything made our managers concerned about not ever, ever overextending ourselves again." Since the early 1970s, EBMUD had been pursuing a large project on the American River to expand its supplies, but legal challenges had stalled the project. At the time of the drought in the late 1980s, the water district was proposing to build a reservoir in Buckhorn Canyon in the East Bay hills that would store water obtained from the American River. Environmentalists and nearby residents' opposition to flooding a local canyon brought water issues to public attention and heightened awareness of the link between water supply and growth. Over the next several years, interests opposing growth took advantage of water's increased salience to challenge several development projects. After defeat of the Buckhorn proposal and the end of the drought, public interest in water issues waned by the mid-1990s.

Rohnert Park, Sonoma County The dispute in the second California region centered not on identifying a supply source and provider for planned growth, but instead on the impacts that increased water consumption might have on neighboring communities. After a decade of local battles over growth, Rohnert Park, a city located north of San Francisco in Sonoma County, embarked in 1998 on a two-year process to update its general plan. The final agreement called for a one percent annual growth rate over the plan's twenty-year life. The municipal water department declared its readiness to serve the development provided for in the plan, but residents of a neighboring community argued that the plan overtaxed Rohnert Park's water supply. Alleging that Rohnert Park was relying on overoptimistic supply projections from the county water wholesaler, the Sonoma County Water Agency (SCWA), the neighbors argued that the city would need to tap a depleted groundwater basin in order to satisfy the increased water demand.

Land-use issues have long been contentious in Sonoma County, where residents seek to maintain the area's rural and agricultural character in the face of substantial growth pressures. Between 1970 and 2000, the county's population more than doubled, and the population of Rohnert Park increased sevenfold to forty-two thousand residents. In 1990, facing a legal challenge from the Sierra Club, Rohnert Park abandoned its right to annex any more territory. One city council member recalls that the "growth wars" that followed over the next decade were "the worst thing that ever happened to this city." Growth was the most salient issue in every city council election, and the council fluctuated between progrowth and antigrowth majorities. According to another local official, in the council elections of the 1990s, "Growth was the issue. I mean, [growth] was a big time issue. There may have been other issues, but I don't remember them."

After the council repeatedly failed to reach consensus on a general plan update that would map out Rohnert Park's future growth, voters took hold of the issue in 1996 and narrowly passed a four-year growth moratorium and urban-growth boundary (UGB). The community reaffirmed its support for the moratorium two years later, defeating by a two-to-one margin a proposal sponsored by the city council to expand and weaken the growth boundary. After the defeat of what critics called "the fake UGB," the council designated representatives from each side of the debate to seek a compromise that could be enacted in a general plan update. Two years of public workshops, focus groups, and drafting produced the plan the council adopted in 2000. As part of the final plan, voters approved a twenty-year UGB, making Rohnert Park the seventh Sonoma County city to adopt such a boundary.

By and large, water issues were peripheral to the county's debates about growth after SCWA overcame environmental opposition and completed construction of a dam on the Russian River in the early 1980s. The statewide drought that occurred later that decade did not seriously jeopardize the county's water supply. However, as new residents continued moving into Sonoma County, attracted by its pastoral charm and relatively inexpensive housing, the county water agency faced supply shortages that kept it from meeting its contracting communities' water demands. In 1998, the agency approved a supply-expansion project that would increase by one-third the amount of water it diverted from the Russian River. The project immediately faced regulatory challenge from federal environmental officials and a legal challenge from environmental, sportfishing, and Native American groups arguing that SCWA's environmental review did not fully address the project's impacts.

These actions stalled the water agency's expansion plan just as demands on the system were peaking. In July 1999, SCWA notified the state health department that its system was within 10 percent of total capacity, and it began urging its contracting water agencies to adopt voluntary conservation measures to reduce their demand. As the severity of the county's water supply problems became more apparent, they attracted growing attention from the media and the public, reaching a peak during the latter stages of the Rohnert Park dispute.

Pennsylvania
Pennsylvania exercises less oversight of local planning activities than California does, creating a less participatory planning process in local communities. Local governments have authority to plan for land use, but municipalities are not required to plan or zone; only counties must adopt a comprehensive plan. Pennsylvania's Municipalities Planning Code includes a number of provisions requiring public notice of land-use actions, but these requirements are limited, and there is no provision for the public to influence land-use policy directly through initiative or referendum. Moreover, the state has no mandate comparable to California's that requires environmental review of development projects.

The smaller set of land-use controls in Pennsylvania reveals the different character of the state's growth challenges. One of the nation's slowest-growing states, Pennsylvania strives to maintain its population, especially in urban areas. But even though many parts of the state have lost population in recent decades, a number of communities on the metropolitan periphery have experienced explosive growth. This imbalance between population growth in some areas and decline in many others has created significant challenges for the distribution of infrastructure and resources; some communities struggle with overcapacity, whereas others must stall new development in order to build facilities for a growing population.

The Pennsylvania cases examined in this chapter are an example of this type of coordination challenge. Their locations are shown in figure 6.3. In both cases, a growing community was forced to slow the pace of development while it sought out new water supplies. Unlike the California disputes, there was never any serious question in these cases

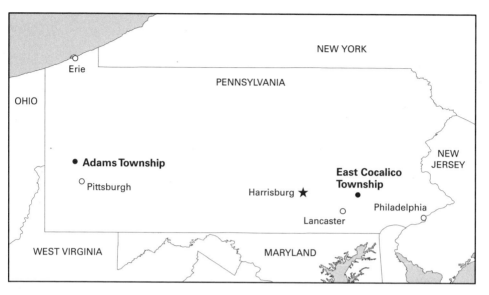

Figure 6.3
Pennsylvania case study locations.

about overall water availability, but securing water resources required establishing agreements with neighboring governments. At the same time, the municipalities making land-use decisions needed to cooperate with the independent water authorities in their jurisdictions to coordinate water and land-use planning. In both Pennsylvania cases, the communities ultimately managed to contract for the water supply needed to support continued development, but not before conflicts between the municipality and the water authority jeopardized negotiations with potential suppliers.

Adams Township Adams Township is located in Butler County, approximately twenty miles north of Pittsburgh in western Pennsylvania. The small township is one of the region's few communities that has experienced growth pressure in recent decades. Whereas the Pittsburgh metropolitan area as a whole has been losing population, Butler County has enjoyed a substantial influx of new residents—its population increased by 15 percent between 1990 and 2000. Two communities in the southwestern corner of the county, Cranberry and Adams, accounted for more than half of this population growth. The population of Adams Township grew by 73 percent during the 1990s, reaching 6,774 in 2000.

This rapid growth in what had been a rural area required Adams to seek out a new water source. The community turned to West View Water Authority, an independent special district that provided water to the booming communities north of Pittsburgh. Before negotiations between the Adams Township Water Authority and its counterpart in West View had reached completion, Adams Township officials requested that the Adams Authority delay signing a contract until the township had finalized its new comprehensive plan and revisions of its zoning law. The township's delay in carrying out its planning process nearly prevented an agreement that would secure a water supply for the benefit of existing and future township residents.

East Cocalico Township The site of the final case is East Cocalico Township, located in the northeast corner of Lancaster County and on the outer rim of the Philadelphia metropolitan area. Like Adams, East Cocalico sought new water sources to accommodate the community's plans for growth. In East Cocalico, however, township supervisors did not wait for the water authority to secure resources before planning and approving new development. The township's aggressive approach to growth brought it into conflict with the local water authority and Lancaster County officials, who insisted that the local water supply and facilities were inadequate to meet the township's ambitious plans.

In contrast with the dispute in Adams, the dispute in East Cocalico took place in a context where growth was highly salient and largely unpopular. A housing boom during the 1980s had caused substantial loss of farmland in this region, which has a rich agricultural heritage. Lancaster County residents are uniquely dedicated to preserving farmland; it is a religious commitment for the county's Amish and other plain sects, and newer residents value the pastoral landscape and the economic benefits from agriculture and tourism. The loss of farmland drew nationwide attention and prompted an ambitious farmland-preservation and growth-planning effort. As of 1997, Lancaster County led the nation in the number of farms it had preserved, and fifteen of the county's municipalities had established UGBs. Lacking enforcement authority, these UGBs did little to control development, but they demonstrate the region's interest in addressing growth issues (Harris 2003).

Although water did not have the same level of salience as it did in California, it attracted more public attention in Lancaster County than in other parts of Pennsylvania. In 1994, the county established a water-resources task force to focus attention on the area's groundwater supply. The task force issued a report in 1997 calling for more wellhead protection to ensure groundwater quality and for intermunicipal cooperation to coordinate water supply. As in western Pennsylvania, the problem was not an overall shortage of water, but rather the distribution of the water supply among providers: the task force estimated that one-third of the county's large water utilities would have insufficient supply to meet 2010 demand, one-third would have enough, and the remaining third would have a surplus. The region surrounding East Cocalico had the most pressing supply concerns, and residents knew it. In a 2001 survey, 82 percent of residents agreed that the region needed to take action to ensure its water supply; support was stronger for action on water than for action on any other issue, including traffic (Lancaster County Planning Commission 2004). For an area with sufficient water resources overall, difficulties with intergovernmental coordination had launched the issue to a prominent position on the public agenda.

Outcomes

All of the disputes described here involved proposals for development that existing water resources could not support. In some cases, water supply issues substantially scaled back growth plans; in other cases, development went forward as originally proposed. Governing structure alone was not decisive in determining the outcomes of these disputes, but divided authority consistently added time and cost to reaching agreement. When coordination depends on bargaining between independent agents rather than within a hierarchical relationship, the costs of negotiation rise. Water agencies want assurance that the city or county will stick to a specified growth plan; land-use officials face an information asymmetry regarding the amount of water that is or might be available. Negotiation becomes even more difficult if either side anticipates that the other will fail to keep its promise after the parties reach agreement (Williamson 1975).

Adding to the costs of reaching agreement is the obligation to carry out negotiations in public. In the unified structure of a general-purpose government, the city or county council can work with planners and water department personnel to find balance between growth imperatives and resource constraints. Special district governance makes it more likely that coordination across issues will take place in public after a proposal has been introduced and framed rather than at the staff level and earlier in the policy process. Negotiating in the public sphere makes it more costly for politicians to compromise and invites new participants into the decision process. With broader participation comes more negotiation, a process that may become more difficult over time as adversarial relationships develop among neighboring governments. Although specialized governance consistently makes coordination more complicated, the dynamics of conflict expansion vary across communities based on the local salience of growth and water issues and on the flexibility of special district boundaries. These contextual factors affect interest groups' ability to mobilize support and seek out venues that would be favorable to their goals.

Dougherty Valley and EBMUD: The Environmental Majority

This is not about water; this is about power. It's about control of land use.
—Dougherty Valley project developer (qtd. in Locke 1993, A1)

We don't have any particular position pro or con on the development itself, but we just don't have the water to supply it.
—EBMUD director (qtd. in Locke 1993, A1)

The complex timeline of events in the Dougherty Valley case appears in figure 6.4. In March 1991, the Contra Costa County Board of Supervisors began considering a proposal by two developers to build eleven thousand homes in Dougherty Valley, what would be the largest residential development project in the county's history. Dougherty Valley fell within the urban limit line passed by voters in 1990, but the county general plan designated the valley as agricultural land. In 1991, it remained undeveloped.

Over the next year and a half, the county carried out its planning and environmental review process under what it called an "aggressive schedule."[4] Water availability was a critical point of uncertainty. Only a small fraction of the nearly 6,000-acre project fell within the EBMUD planned ultimate service boundary; the rest lay outside the boundaries of any water utility. The draft environmental review released in June 1992 concluded that the development could be annexed to EBMUD or to a neighboring water utility, the Dublin San Ramon Services District (DSRSD). The draft specific plan for the location assumed that EBMUD would be the primary water provider.

Fighting over Land and Water: Venues in Local Growth Disputes 141

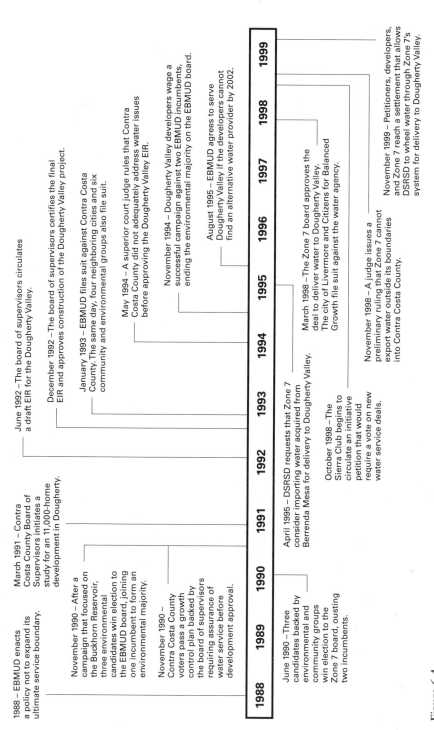

Figure 6.4
Timeline of Dougherty Valley events.

Despite the county's expectations, EBMUD insisted throughout this planning period that it lacked sufficient water supply to serve Dougherty Valley. The district already had been suffering supply shortages when it enacted a policy in 1988 to freeze its service boundary. Drought conditions and regulatory mandates added to the pressure on EBMUD's water resources, and the district's supply outlook appeared even dimmer after a 1989 court ruling put a halt to EBMUD's expansion plan. From the time of the project's initial proposal, EBMUD sent letters insisting that it would not expand its service area when it could not guarantee reliable water service even to its existing 1.2 million customers. The water district reiterated this position in formal comments responding to the project's environmental documents. The county ignored these warnings, however, naming EBMUD as the likely service provider to Dougherty Valley in the final EIR released in November 1992.

Contra Costa County's failure to negotiate early with EBMUD or to find an alternative water provider pushed the problem of water into the public sphere. The well-publicized conflict between the two governments drew attention to uncertainty about water supply for the valley's development. Water issues became an important part of the larger debate over the proposal, with environmental and community groups, area residents, and the county health department raising concerns about water availability in their comments on the draft environmental documents. Media coverage also consistently cited water availability as a leading issue in the public discussion about valley development. Participants perceived the threat to water resources as real, and the threat was particularly meaningful because of the drought that had recently ended. However, EBMUD's public resistance to providing water service also provided ammunition for groups and neighboring communities that already opposed the project because of its potential impacts on traffic and quality of life in the region. California's environmental review process provided a forum for the neighbors of Dougherty Valley to express their concerns about the development proposal, but their arguments were more forceful because they drew on EBMUD's assertion that providing water to the valley would threaten service to existing customers.

The public nature of the dispute served to solidify both governments' positions. Despite challenges from EBMUD and other neighboring governments that the project's EIR still failed to provide sufficient analysis of water impacts, the county board of supervisors certified the EIR and approved the Dougherty Valley project in late December 1992. The schedule allowed the longtime supervisor who represented Dougherty Valley to vote in favor of the development before stepping down from office; the candidate elected a month earlier to take his place had strongly opposed the project during her campaign.

EBMUD immediately filed suit, arguing that the county had shirked its duty to identify a legitimate water source:

> Most of Dougherty Valley is not only outside EBMUD's existing service area, it is outside EBMUD's planned service boundaries. Nonetheless, the County cavalierly designated EBMUD as the primary water provider in the face of specific and uncontroverted evidence that EBMUD does not have adequate water in the foreseeable future to serve the project. The County's actions can only be based on some unfounded faith that with the project approval, water will come.[5]

The district also began pursuing legislation that would give water agencies statewide a role in approving new development.[6] The county countersued and introduced its own bill in the state legislature requiring the district to provide service to Dougherty Valley. Moreover, by this time, the conflict no longer was limited to the county and the water district. Five neighboring communities teamed with several environmental groups to file suit against Contra Costa County, contesting the environmental review, and their brief repeatedly referred to EBMUD's argument about its ability to serve existing customers.[7]

Whereas neighboring governments cited EBMUD's stance as evidence of real resource limits in the region, the county and project developers argued that the water district had arbitrarily restricted its water supply with the goal of blocking growth. Feeding their suspicions was environmental groups' recent success in electing a majority to the nonpartisan EBMUD board. The Sierra Club and other environmental groups in the region had been confronting the water district since the 1980s over proposed water-storage facilities that would allow EBMUD to serve growing communities. The groups relied on outside strategies to influence the district until 1990, when the planned construction of a new storage reservoir and anticipation of development proposals in Dougherty Valley prompted the groups to run a slate of three candidates for the water district board. The central issue in the election was growth. According to a board incumbent challenged by one of the environmental candidates, "If [the reservoir] isn't built there won't be any new water development, and badly needed housing and new jobs will be seriously jeopardized." The challenger in another ward characterized the reservoir project as "basically a gift of ratepayers' money to developers so they can convert agricultural land to suburban developments" (both quotes in Halstuk 1990, 2).

The county board of supervisors had direct authority over the development decisions that environmentalists sought to influence, but the groups expected to have more success achieving their policy goals by concentrating on EBMUD. One reason was the lesser expense of a campaign for EBMUD office. Boundary design created another advantage, because EBMUD overlapped into the politically liberal areas of neighboring Alameda County. Developers, too, became involved in the 1990 election, contributing to the environmentalist's opponent in each race. These efforts backfired, however, because unusually high levels of press and public attention to the election allowed the environmental candidates to depict their opponents as captive to development interests. After a contentious campaign, all three challengers narrowly won election, and they joined an incumbent environmentalist to form a four-vote majority on the seven-member EBMUD board.

Once in office, the new board carried through on campaign promises to change direction at EBMUD. It abandoned the reservoir project and developed a new plan for water supply management that relied on conservation and conjunctive-use groundwater storage. The board adopted an increasing block rate structure and then spent several years in court defending the rates against a lawsuit brought by residents in Contra Costa County, who had bigger lots and warmer summers than in the urban Alameda County sections of the district's territory. EBMUD also reaffirmed existing policy not to expand the district's service boundary, sparking the long dispute between the district and Contra Costa County.

Environmentalists' success in winning control of the EBMUD board owed much to the high salience of water issues in that region during the early 1990s. The 1990 election took place during a severe statewide drought, and those who sought a new direction in local water policy took advantage of heightened public concern about water scarcity. The groups targeted EBMUD wards located in liberal sections of Alameda County, far from Dougherty and Tassajara valleys, and focused the campaign on the flooding of a scenic canyon accessible from Alameda County communities. Responding to what it perceived as a strong environmental, antigrowth message from voters, the water board took an aggressive stance against the Dougherty development by suing Contra Costa County. Were it not for the drought and the subsequent election of environmental board members, an EBMUD staff member said in an interview, the water district would have resisted serving Dougherty Valley, but it would not have filed the lawsuit. Instead, he explained, "We would have done what every other water agency has done—we would have bit the bullet."

Responding to these policy changes, Dougherty Valley project developers attacked EBMUD in the press and public statements, accusing the district's board of directors of intentionally restricting water supply in order to pursue an antigrowth agenda. They hired an investigator to audit EBMUD's finances and released a report to the press detailing wasteful spending by the district, including $29,000 spent on bottled water. They also pursued an electoral strategy, backing a challenger to the one member of the board's environmental majority facing reelection in 1992. The incumbent kept her seat, maintaining the environmental majority for two more years.

By 1994, the drought had ended, the canyon reservoir project was no longer under consideration, and the salience of water issues had declined. Project developers worked with the Building Industries Association to wage a well-funded campaign against the environmental board members running for reelection. Environmental groups backed the incumbents, but the board elections received little attention from the media, and the incumbents had limited resources for responding to the campaign against them. Efforts focused on the environmental incumbents in two swing wards, one located entirely within Alameda County and the other predominantly within Alameda's boundaries. The election was a referendum on EBMUD's growth policies, and voters expressed disapproval of the district's aggressive stance. Both environmental incumbents were defeated, and the EBMUD board came out of the election with what one director called a "5-to-2 pragmatic majority" (Haeseler 1995, A1).

Although they ran a vigorous grassroots campaign, the environmental groups were unable to sustain the interest and participation of voters in liberal parts of Alameda County who had delivered narrow victories to environmental candidates four years earlier. Table 6.1 shows turnout figures in the four wards electing board members in 1990 and 1994. Across all three wards that overlapped with Alameda County, participation among Alameda voters declined. Although the county's overall turnout in the election for governor and other statewide officials increased by 9 percent, the declining salience of water issues for Alameda residents produced a higher rate of abstention in the EBMUD races than four years earlier. Wards 3 and 7 were the two swing precincts. In Ward 7, located

Table 6.1
Turnout in EBMUD Elections, 1990 and 1994

	Votes cast					
	Alameda County			Contra Costa County		
	1990	1994	Change	1990	1994	Change
Ward 2*	—	—	—	60,247	47,282	−22%
Ward 3	42,491	40,296	−5%	10,018	14,533	+45%
Ward 4	38,614	35,743	−7%	9,879	13,773	+39%
Ward 7	40,422	37,960	−6%	—	—	—
Countywide, all elections	381,038	414,302	+9%	277,999	305,312	+10%

Note: * The incumbent in Ward 2 ran unopposed in 1994.
Sources: "Alameda County" 1994; "Complete Results" 1990; "Contra Costa County" 1994; Eu 1990; B. Jones 1994.

entirely in Alameda County, the environmental candidate won a narrow victory over an incumbent in 1990 and then lost to a developer-backed candidate when turnout declined four years later. The environmental candidate in Ward 3 maintained the same level of majority support among Alameda voters in both elections, but the decline in turnout within Alameda and the increase in Contra Costa County changed his narrow victory in 1990 to a narrow defeat in 1994.

By 1994, water supply was no longer a severe problem in Alameda County, so the public became less attentive to EBMUD elections. Antigrowth interests could not take advantage of water salience to rally public opposition to proposed development in a neighboring county. Within Contra Costa County, controversy over the growth implications of the Dougherty Valley development was reaching its peak in 1994. Public opinion was divided, but the growing salience of development issues among residents on both sides of the dispute produced higher levels of participation in EBMUD elections. For the two seats contested in Contra Costa County, turnout among county voters increased by approximately 40 percent between 1990 and 1994.[8] The changing salience of water and growth issues heightened attention to EBMUD in the more conservative portions of the water district's jurisdiction, but caused it to fall in the liberal areas, creating conditions for a change in control of the EBMUD board.

After the 1994 election, the EBMUD board sought to work more cooperatively with Contra Costa County. Within months, it reached a legal settlement with the county, agreeing to annex and serve Dougherty Valley if the project's developers could not find an alternative water provider by 2002. The water district's new leadership sought to establish a less antagonistic relationship with the county, and because water issues were attracting less public attention than they were a few years earlier, the EBMUD board seemed to have district constituents' support. The settlement did not resolve the question of water service for Dougherty Valley, however, but merely transferred the conflict to a new policy venue.

Dougherty Valley and Zone 7: The Shadow Agency

The bottom line was it was in the best interests of the residents—that they were going to have increased reliability, increased water resources, increased groundwater storage, and a drop in rates of 8 percent. So, why was I elected? To represent the best interests of the current residents.
—Zone 7 Water Agency board member (personal interview)

Zone 7 defines antidemocratic practice.
—City of Livermore and Citizens for Balanced Growth[9]

The settlement with EBMUD sent Dougherty Valley's developers in pursuit of another water source for their project. They soon struck a deal to purchase 7,000 acre-feet annually from the Berrenda Mesa Water District in faraway Kern County. The size of the purchase well exceeded the approximately 6,000 acre-feet required in Dougherty Valley, but the challenge was to find a way to deliver the water to the new homes.

The DSRSD was anxious to be Dougherty Valley's water utility. Incorporated in the 1980s, the City of Dublin sought to expand. The separately elected services district providing Dublin's water and wastewater services shared this progrowth vision. Annexing Dougherty Valley would double DSRSD's service territory, and the district believed that adding additional customers would reduce water rates systemwide (Farooq 2005). DSRSD could not act on its own, however, because the district lacked conveyance rights and facilities to import the Berrenda Mesa water. Moreover, by contract it could not expand without the approval of its wholesale water supplier, the Alameda County Flood Control and Water Conservation District, commonly known as Zone 7.

Prior to 1990, Zone 7 had been what a subsequent board member calls a "shadow agency.... Nobody had ever heard of it." As a wholesaler, Zone 7 did not sell water directly to its constituents; it supplied treated water to DSRSD, the cities of Pleasanton and Livermore, and a private water company serving the Livermore area, and it sold untreated water to agricultural customers. There had historically been little competition for seats on the board, but during the drought in 1990 a local group worked with members of the Livermore City Council to put together a slate of environmental candidates who raised questions about water planning and water quality. Zone 7's water supply came primarily from the State Water Project, which had contracts to deliver more water statewide than the system could provide. The environmentalists running for the Zone 7 board argued that the agency could not rely on receiving its full contracted allotment from the state project when planning for growth. They also called attention to the fact that Zone 7's water was not meeting federal drinking water standards. Three of the four candidates on the slate won election.

In February 1991, less than a year after the new board took office, Zone 7 heard from the State Water Project that drought conditions had forced cutbacks that would result in the agency's receiving just 10 percent of its contracted water for the year. Zone 7 had groundwater supplies it could use in emergency conditions, but the announcement prompted the agency to take a more aggressive approach to water planning. Two Zone 7 directors proposed a "fair share" policy that would set a cap on annual water deliveries; development projects requiring a larger allocation would go to a vote of Zone 7's electorate. The proposal met strong opposition from the agency's contracting cities and developers, who argued that Zone 7, like EBMUD, was attempting to use water supply as a tool to control growth. Proponents of the plan maintained that the agency faced real supply constraints and that residents should have an opportunity to express whether they wanted to pay higher rates for the system development necessary to serve new growth. As one of the directors who authored the proposal explained at the time,

> There are two very different views of what Zone 7 should be. On the one hand, some people believe it should be a utility—providing as much water as anyone wants, no matter the cost. People at the other extreme view the agency as the ultimate growth management tool. In reality it's neither. I think we should educate people that there is a finite amount to this resource. Then there are two options: we can stop developing, or we can spend more money for more water after getting an approval from voters. (qtd. in A. Miller 1992, 1)

Fighting over Land and Water: Venues in Local Growth Disputes 149

The strongest opposition came from Dublin and the DSRSD, who threatened to seek out another water supply source if Zone 7 approved the proposal (Brazil 1992; Saltonstall 1992; Vonheeder and McCormick 1992). Facing resistance from its customer utilities, Zone 7 abandoned the plan and began to enact a series of rate hikes to cover the cost of supplementing its supply.

These events were unfolding while Contra Costa County was preparing its EIR for the Dougherty Valley proposal, and they kept the county from seriously considering DSRSD as a potential water supplier. DSRSD sought to expand into the valley, but the need to secure water rights and Zone 7 cooperation posed too great an obstacle. By the time EBMUD settled with the county in 1995, however, the situation had changed. The drought had ended, increasing the region's water supply and reducing the salience of water issues. Even more important, Dougherty Valley's developers had obtained water for the project and needed only a means to transport it.

In April 1995, DSRSD requested that Zone 7 import the water that the Dougherty Valley developers had acquired from Berrenda Mesa. The Zone 7 board spent the next year studying and considering the request. It consulted with Livermore and Pleasanton, contracting cities in which many local officials had opposed the Dougherty Valley development, but neither city raised objections to Zone 7's involvement in importing water for the project.[10] In June 1996, the Zone 7 board directed staff to prepare agreements necessary for it to provide Berrenda Mesa water to Dougherty Valley. Nearly two years later, after extensive studies and environmental review, the board approved the deal.

Zone 7's board of directors attributed its decision to the benefits the arrangement offered for current residents within the agency's jurisdiction. For agreeing to wheel the water that Dougherty Valley developers had acquired through Zone 7 pipes and send it to the DSRSD, Zone 7 would obtain a new water supply that would improve reliability for its existing customers, greater flexibility in managing groundwater resources, and as much as $100 million in savings to current residents from connection fees and a larger customer base. Moreover, the board recognized that the development project would proceed even without Zone 7 involvement. EBMUD had committed to providing water to Dougherty Valley after 2002 if the developers found no other alternative. DSRSD was threatening to establish its own conveyance rights and facilities to serve Dougherty Valley alone, perhaps exercising its buyout option and dropping Zone 7 altogether (personal interviews; Vonderbrueggen 1996; Zone 7 Water Agency 1997).

Zone 7's agreement to cooperate with DSRSD came over the opposition of groups opposing the Dougherty Valley development. Antigrowth activists had lobbied Zone 7 not to approve the arrangement with DSRSD, and the agency's decision drew substantial public criticism and opposition. One Zone 7 director recollects, "On that Dougherty Valley vote, I lost probably 50 percent of my core political support.... And people that had been long-time supporters were publicly calling for me to be recalled from the Board." After the decision, the local Sierra Club joined with another community organization to collect signatures for a ballot initiative requiring a public vote for service-expansion proposals. These groups worried that the Dougherty Valley deal would set a precedent for Zone 7 providing water to future projects in undeveloped Tassajara Valley and other neighboring areas. They also hoped that litigation would hold up the agreement until after their initiative passed, making the agreement subject to voter approval. Despite these hopes, a judge ruled that any initiative would have to go before all voters in Alameda County, not just those residing within Zone 7's jurisdiction, thus substantially raising the burden for signature collection. The groups ultimately dropped the effort.

In addition, a community group called Citizens for Balanced Growth joined with the City of Livermore to challenge the Zone 7 water deal in court. In their brief, the petitioners argued that Zone 7 "defined antidemocratic practice" by violating its responsibility as trustee to its constituents and failing to protect their scarce water supply.[11] Ironically, although opponents to the Dougherty Valley project had chosen to pursue their agenda through appeals to water districts, at the same time they were making legal arguments that these districts failed to meet minimal standards of democratic responsiveness. In fact, the lawsuit was part of a political strategy to return the Dougherty Valley development to EBMUD, which had a legal obligation to provide water to the project if developers could not acquire it elsewhere.

Antigrowth groups fighting Zone 7 in court and with ballot initiative threats seemed to be anticipating the next big proposal in Tassajara Valley. If the project returned to EBMUD, environmentalists thought they could make board members pay an electoral price for the legal settlement, allowing election of a new board that would be less willing to expand when future project proposals came forward. In addition, although Livermore City Council members raised no objections when Zone 7 consulted them about the Dougherty Valley deal, the city later filed suit as a means to keep Zone 7 from providing water service to pending development projects (Vonderbrueggen 1998). The parties eventually reached a settlement that allowed DSRSD to wheel the Dougherty Valley water through Zone 7's system but restricted DSRSD's ability to expand. In November 1999, seven years after winning rapid approval for the largest housing development in the history of Contra Costa County, Dougherty Valley's developers had secured their water. But groups opposing growth in the region were successful in making sure that for the forthcoming Tassajara Valley proposal, the only option for water supply would be EBMUD.

Where separate authorities are responsible for decision making about land use and the infrastructure demanded by new development, proposed development projects face more potential veto points. Water district officials have the power to prevent and direct community growth through their control over water supply and the pipes required to deliver it. Specialized governance therefore offers antigrowth interests an opportunity to focus their efforts on the most favorable policy venue. In some cases, this venue may be the general-purpose land-use authority; in the East Bay, opponents of Dougherty Valley development chose to concentrate attention on water districts.

Although environmental and community groups joined in a lawsuit challenging county approval of the project, they never made a serious attempt to influence the Contra Costa County Board of Supervisors' decision making. Instead, they pursued an explicit strategy to challenge growth at EBMUD, which they perceived would be more favorable to their interests. The perception stemmed not from the district's limited functional scope, but from its boundaries, which encompassed a more liberal constituency than the county—thus allowing election of directors who shared the environmentalists' position against sprawling development. The strategy to target EBMUD was successful for a time, until the salience of water issues declined and Alameda County voters lost interest in the land-use issues of a neighboring county. At that point, activists seeking to block the project turned their attention to Zone 7, historically the type of shadow agency described by critics of special districts. Although Zone 7 had attempted to influence growth policy a few years earlier, the high salience of the Dougherty Valley dispute ultimately led the agency to cooperate with the county's development plans, just as

EBMUD was ready to do. The strategy used by environmental and community groups strategy was not entirely unsuccessful, however: by electing a majority to the water district board, they prevented EBMUD from dedicating a significant part of its limited water supply to a large new development in Dougherty Valley, and their efforts in the Zone 7 venue prevented DSRSD from expanding its jurisdiction beyond Dougherty.

The existence of additional veto points in a fragmented system provides opportunity for those who oppose growth to block a development proposal, but the more flexible boundaries of overlapping special districts allow developers to shop for services among multiple potential providers. Services from a consolidated city or county come as a package; even where a choice might exist over which city will annex a new subdivision in an unincorporated area, the developer must find a city that can provide all the services the project requires. Specialized governance divides up the bundle of urban services, allowing developers who cannot reach agreement with one special district to shop for the service among alternative providers.

The developers behind the Dougherty Valley project first sought to obtain water from EBMUD, a large water district whose jurisdictional boundaries lay closest to the valley. Meeting resistance from EBMUD, the developers pursued a dual strategy: they put political pressure on the district to reverse its opposition, and they began looking for an alternative supplier. They obtained supply rights from a water district located 250 miles away, and they started negotiating with DSRSD and Zone 7 to transport the water to Dougherty Valley. The developers' efforts in multiple venues ultimately ensured the provision of water service. Knowing that the developers were pursuing alternatives with other water districts, EBMUD consented to serve the valley in the unlikely case that the other options did not work out. Moreover, the agreement with EBMUD was a major factor in Zone 7's decision, according to a Zone 7 director: "The way I explain this to people is if East Bay MUD supplies this project, we will have the increased air quality impacts, the increased traffic impacts on this hand. On this hand, we have the increased air quality impacts, traffic, and $100 million." By appealing to multiple providers, the Dougherty Valley developers were able to obtain the water supply they needed. In the end, specialized water governance did not influence the land-use outcome of the Dougherty Valley dispute—the project went forward as originally proposed, as it almost surely would have done with consolidated governance of land use and water. Specialization did greatly increase the project's cost to its developers and set the stage for the next major housing proposal in Contra Costa County.

Tassajara Valley, Round 1: The Defeat

We want to draw the line in the sand early on.
—EBMUD director (qtd. in Hallissy 1997, A13)

Figure 6.5 shows the timeline for the Tassajara Valley case. Soon after the Contra Costa County Board of Supervisors approved development in Dougherty Valley, landowners in nearby Tassajara Valley began developing a proposal for more housing there. Like its neighbor, Tassajara Valley consisted of predominantly undeveloped agricultural land located in an unincorporated area of the county. Approximately twenty valley landowners had joined together as the Tassajara Valley Property Owners' Association (TVPOA) to work on a development strategy for the area.

Coming on the heels of the county's controversial approval of the Dougherty Valley project and during the extensive litigation over that project, the idea of further growth into the next valley drew fierce opposition. Tassajara Valley became a central issue in the 1996 candidate races for the EBMUD board and the county board of supervisors. The election of three new supervisors to the five-member county board gave confidence to antigrowth activists that they would have a better chance to defeat proposals to develop Tassajara Valley, even though it was not clear that a majority of the new board was on their side (Hytha 1997).

In March 1997, the county released the EIR for a six-thousand-home development on 4,500 acres in Tassajara Valley. Within weeks, the EBMUD board voted unanimously not to provide water. One director commented, "It's done to warn them in advance and early on that we do not have enough water for the project. We want to draw the line in the sand early on" (qtd. in Hallissy 1997, A13). The EIR named DSRSD as well as EBMUD as potential providers, but even the Dublin-based district responded that it lacked adequate supply to serve the project. The TVPOA said that it was willing to find its own supply, but that it would need cooperation from a water utility to transport the water to the valley: "We're perfectly willing to find our own water. There's plenty of water out there. The issue is not supply. It's being able to convey it" (qtd. in Hallissy 1997, A13).

154   Chapter 6

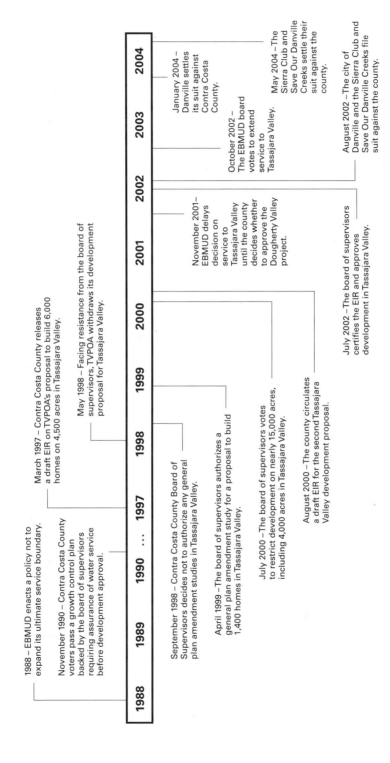

Figure 6.5
Timeline of Tassajara Valley events.

The TVPOA withdrew its development proposal in May 1998, however, just before the county board of supervisors was to consider it for approval. The property owners group said that antigrowth sentiment had prompted the withdrawal; observers attributed the decision to uncertainty about water availability (Hytha 1998; Jacobus and Hytha 1998; King 2001). In September, county supervisors voted not to authorize any general plan amendment studies in Tassajara Valley until they carried out a major countywide assessment of land-use policy. Dougherty Valley had raised the salience of growth issues even more than was typical in the region, and supervisors were not prepared to approve what would amount to another new community before the dispute over Dougherty Valley had been resolved.

Tassajara Valley, Round 2: The Bargain

It was political gamesmanship in the San Ramon Valley.
—EBMUD director (qtd. in Davis 2002, A3)

In April 1999, despite its recent commitment not to conduct such studies, the Contra Costa County Board of Supervisors authorized a study on a proposal to build 1,400 homes in Tassajara Valley. Sponsors of this proposal included a major development company that also was involved in the Dougherty Valley project. Widely considered an influential political actor in Contra Costa County, the company had not been part of the TVPOA.

Release of the draft EIR for this new Tassajara Valley proposal in August 2000 sparked another public dispute between the county and EBMUD. Approximately half the development fell within EBMUD's ultimate service boundary, but 729 proposed homes lay outside the jurisdiction of any water agency. The county again assumed that EBMUD would provide service to the entire project, despite the water district's assertions to the contrary. The district sent letters to the county board in 1999 and 2000, stating that annexing new service territory would increase the severity and frequency of water shortages for existing customers.[12] In its comments on the project's environmental review, EBMUD objected that the county had ignored the district's earlier letters and relied on "obsolete and erroneous" information. The district specifically rejected the county's proposal that developers fund conservation programs to offset the project's water consumption, arguing that it already intended to implement the same conservation measures in order to meet existing demand. Ignoring EBMUD's opposition, the final EIR released in September stated that "EBMUD is the logical service provider" for Tassajara Valley.[13]

Notwithstanding its firm public stance, the EBMUD board maintained some flexibility on supplying water to the Tassajara development. Rather than take an early formal position on serving the project as it had for Dougherty Valley and the first Tassajara proposal, EBMUD decided to delay its decision until the county voted to approve the project. The county took more time considering this project than it had in the case of Dougherty Valley, a project eight times the size of the pending Tassajara proposal. It conducted hearings and negotiations throughout 2001 and the first half of 2002, and it began negotiating with EBMUD. The two governments ultimately reached an agreement that EBMUD would supply water for the development in exchange for an $8.5 million conservation investment by developers that would result in two gallons of water saved for every gallon consumed in Tassajara Valley. In July 2002, county supervisors approved the project, and the EBMUD board followed in October with a four-to-three vote to accept the conservation plan and annex the full Tassajara Valley development into its service area.

Although the county and the water district were eventually able to reach compromise on the Tassajara Valley project, the two years they spent in conflict contributed to public opposition to the project. Residents of nearby cities as well as environmental and community groups consistently voiced their objections to the development, citing concerns that the project would jeopardize water supplies to EBMUD's existing customer base (Johnson 2002). EBMUD directors knew that their ultimate decision to supply water to the valley would be controversial. A director who had endorsed serving the project from early on argued that the district was backed into a corner: "I realize that it's a very politically charged issue and that I will lose votes because of it. But if we had acted differently, a judge could have ruled we acted arbitrarily and capriciously and not in good faith. The district would lose in a court of law at that point and we would not be able to mandate any conservation on the project that would have a negative impact on our customers" (qtd. in I. Miller 2002, A1).

EBMUD directors maintained that the project would go forward no matter what decision they made, and they believed they were acting in their customers' best interest given that constraint. Critics charged that EBMUD had "caved" and that the conservation measures would make reducing water consumption more difficult in the next drought.

Project opponents used EBMUD's initial resistance as evidence to support their lawsuits against the county. Consolidated lawsuits filed by the Town of Danville and two community groups argued that the project's EIR failed to analyze adequately the impacts on water supply, traffic, and the valley's steep slopes. Danville's brief suggested in its discussion of water issues that EBMUD had been bought off: "After refusing to extend its [ultimate service boundary] for forty years and vigorously fighting specific Camino Tassajara development projects for over a decade, EBMUD backed down shortly before the County's approval of the Project.... EBMUD's well-compensated change of heart does nothing to alter the fundamental realities of the Project."[14] In short, Danville argued, there was nothing to refute EBMUD's initial concern that supplying the project would exacerbate the district's water shortage and harm its customers. The environmental groups hoped to use their lawsuit to put political pressure on the EBMUD board. One participant recalls, "We were hoping to hold their feet to the fire and maybe get enough political pressure on the board that they would... pull back and say, 'Well, actually, I guess... we don't really have the water to supply this.'" In its responding brief, the county appealed to EBMUD's national reputation as a professional and innovative water district. Whereas Contra Costa County had called into question the EBMUD board's judgment in the Dougherty Valley case, this time it lauded the water district's exacting standards and argued that the petitioners should defer to the district's assessment in approving service.[15] The lawsuits did not reach settlement until 2004, five years after the project was proposed.

The protracted legal dispute over Dougherty Valley kept Contra Costa County from pursuing the first, larger Tassajara development proposal. Growth opponents could not prevent development in the valley entirely, but their efforts produced a smaller project and an impressive water-conservation commitment from Tassajara's developers—a commitment that environmentalists throughout California later attempted to replicate in other development proposals. The opposition also created costs and delay that might have provided a disincentive for developers to pursue other projects. The attorney for one of the development companies remarked about water district review: "It adds another opportunity to delay. In the case of Dougherty Valley, it delayed it for years. Those are the years when the housing doesn't get built. They are the years that add to the housing affordability crisis here in the Bay Area by overloading the housing approval process with all of these hoops that are so redundant that they give opportunity after opportunity to derail projects" (qtd. in Levine 2000, 1).

In moving consideration of water issues into the public sphere, thus increasing the number of stakeholders involved in decision making and creating multiple venues for debate over water supply, specialized governance can slow the path to decision making. As the Pennsylvania case presented in the next section reveals, delay can sometimes put a community's water supply in jeopardy. Yet expanded participation also creates an opportunity for more innovative problem solving—for example, facilitating a long-distance water transfer rather than overtapping local resources in the case of Dougherty Valley or creating a new standard for water conservation in the Tassajara Valley development case. The challenge in a fragmented system is to promote coordination at an early state in policymaking, exploiting the potential benefits of broad participation, but avoiding costly antagonism. Higher authorities can play a role in encouraging this kind of coordination, as California eventually did through the water-adequacy law it passed as a result of the Contra Costa County disputes.

Adams Township: A New Authority

If you turn this water over to the developers, you're giving them a blank check to do whatever they will.
—Adams Township supervisor candidate (qtd. in Weiskind 1995b, N5)

Specialized governance can contribute to escalation of disputes over land use and water even where water is more plentiful. In the early 1990s, new construction in the western Pennsylvania community of Adams Township caused the water table to drop. Township officials set up an independent water authority with the purpose of seeking out and acquiring new water sources. The Adams authority was under pressure to boost its water supply quickly because shortages were preventing additional connections to the system (Weiskind 1995a), so it entered into negotiations with West View Water Authority, the dominant regional water provider. Before the two authorities could reach final agreement,

Adams Township planners and supervisors intervened to request that no contract be signed until September 1995, when the township would complete drafting a new set of zoning laws.

In a consolidated system, a township board would simply prohibit its water department from pursuing a new contract until the board was ready to approve the deal. With divided governance, however, the township publicly appealed to the water authority, attracting community residents' attention at the same time. A group of Adams residents soon thereafter submitted a petition calling for further delay on the supply contract until the new zoning laws could be implemented. The petition backers argued that a water contract without zoning laws in place would allow developers to decide the future of the township. Much of Adams's appeal derives from its rural character, and residents worried that expansion of the town's water supply would erode that appeal. During debate over the new zoning laws, one township supervisor observed, "It doesn't take a rocket scientist to know that if we sign a water agreement (now) the township will explode" (qtd. Natale 1996, NW2, parentheses in the original). The water authority reluctantly agreed to the delay, but the authority chair complained that the township was sending mixed signals: "I don't understand why two or three years ago you asked us to pursue water and now you're asking us to stop" (qtd. in Weiskind 1995c, NW3).

Over the next year, the township continued postponing completion of the new zoning laws. In the spirit of cooperating with the township board, the water authority did not pursue the West View contract, and in May 1997 West View withdrew its offer to sell surplus water to Adams. As West View's director said of the negotiations with Adams at that time, "We haven't heard from them in three years" (qtd. in Natale 1997a, N4). Revealing the confused lines of authority over these complex issues, the chair of the Adams water authority told a reporter he had not heard that West View retracted the offer (Natale 1997a). West View's action raised the stakes on the township's delay, and township board members eventually consented to the authority's renegotiating the West View contract, provided that any extension of water service strictly follow the township's comprehensive plan (Natale 1997b). The township board also hired consultants to help it finish the zoning. After months of uncertainty, the two water authorities renegotiated a twenty-five-year contract to provide Adams with treated water. Adams also began to explore options for developing a new supply source of its own, because West View could not commit to supplying enough water to serve Adams's plans for growth.

The township and the water authority finally agreed about the importance of pursuing a new water source, but conflict over growth goals nearly cost them access to the best available supply option. Divided governance bears some responsibility for the close call. The township did not trust the water authority to follow the guidance of the comprehensive plan in extending water service to new customers, so it sought to use zoning laws to limit the authority's flexibility. By requesting in public that the authority delay signing a supply contract, the township attracted the attention of residents concerned about growth, and their petition and ongoing pressure served to harden the township's position.

The Adams Township Water Authority's experiences demonstrate the tension between the values guiding special district formation and the ongoing pressures to respond to constituents and other governments. Township supervisors first set the authority on a mission to acquire new sources, but later sought to delay signing a water contract. Water-authority directors had invested time negotiating with West View, and they knew that the opportunity to purchase West View's surplus water might pass; two years earlier, West View had refused to provide water to Adams Township, citing concerns about the adequacy of its own supply. The Adams authority was reluctant to agree to the township supervisors' request for delay when it had nearly succeeded in carrying out its initial mission. The authority also began to part ways with the supervisors over beneficiaries of the expanding water system. Adams supervisors initially planned a developer-driven expansion of the water system, with developers funding construction of the system's infrastructure. Once the township's water authority had been established, however, it came under pressure from constituents to provide public water service to existing residents, an endeavor that would require more public spending and a larger role for the authority than originally envisioned. Water-authority officials were ready to embrace this role, even though it departed from the authority's original mission. Those who design institutions attempt to embed their values and goals. Once a special district has been established, however, district officials have more incentive to respond to their constituents than to the politicians that enabled district formation. Like other public officials, district leaders have an interest in protecting and expanding their jurisdiction.

East Cocalico Township: A Progrowth Township

Municipalities are not permitted to say we don't have water. We'd spend a lot of money fighting it in court, and we'd lose.
—East Cocalico Township supervisor (qtd. in Hernon 2003b, B6)

If you know where to find water, let me know.
—East Cocalico Water Authority chairperson (qtd. in Hernon 2003a, B6)

Across the state from Adams Township, public officials in Lancaster County faced a similar dilemma a few years later when booming growth in the county's northeast corner put stress on local resources and infrastructure and raised concerns about preservation of the region's rural character. Located at an important highway intersection, East Cocalico Township became the focus of much of the region's development pressure. Unlike the Adams Township supervisors, the East Cocalico supervisors did not seek to slow the pace of development, but instead pursued an aggressive growth plan. Their attempts to accommodate new housing and commercial development brought the township into repeated conflict with Lancaster County officials, in particular the county planning commission, which had been undertaking a large-scale effort to coordinate planning for the county's groundwater supply. East Cocalico repeatedly approved development proposals over the opposition of county planning officials, who wanted municipalities to give greater attention to water supply when planning for land use.

The conflict reached its peak in 2003 when the township proposed to rezone nearly 750 acres of farmland to residential development. The rezoning would provide for eighteen thousand new residents over the next twenty years, nearly triple the number that the county had anticipated in its own planning process. The growth required an additional 340,000 gallons of water per day, and the township had no plan for obtaining new supply. Lancaster County officials strongly opposed the township's rezoning proposal, as did surrounding townships, the East Cocalico Planning Commission, and many township residents. Perhaps most persuasive was the opposition announced by the East Cocalico Water Authority, which stated in a letter that it would not have the capacity to provide water and sewage facilities to accommodate the planned growth. The authority reported that it already faced a supply shortfall of 150,000 gallons per day (Hernon 2003c).

Similar to what happened in California's East Bay, resistance from the local water district shifted the planning dispute into the public sphere and lent credibility to the arguments presented by citizens who already had opposed the township's growth agenda. In a 2001 survey of the Cocalico region conducted by the county planning commission, residents expressed overwhelming support for slowing the pace of growth: 74 percent said the current growth rate was too fast or much too fast, and 79 percent said that the future growth rate should be slower than the existing rate (Lancaster County Planning Commission 2004, 23-2). Despite opposition from residents and surrounding jurisdictions, township officials sought to continue building. The local water authority's resistance to the township's growth plan bolstered the argument that local resources could not support continued booming development. It also changed the focus of the debate from traditional concerns about farmland preservation to the status of local water supplies. At the public hearing on East Cocalico's proposal, water was the leading focus of residents' concerns (Hernon 2003d).

East Cocalico Township ultimately scaled back its growth plan in the face of water supply limitations, and by late 2003 the township authority reached an agreement with a neighboring community to obtain enough water to support a modest level of growth. Because the water authority had been visibly engaged in efforts to obtain new supply by drilling test wells and negotiating with a neighboring authority, township officials could not allege—as Contra Costa County supervisors did—that the water shortage was artificial. They eventually needed to adjust their growth plans to acknowledge real limitations in supply. Given the real supply shortages, consolidated governance of land use and water probably would not have changed the outcome in this case, but divided governance allowed the conflict to escalate. Seeking to avoid accountability for their unpopular progrowth position, East Cocalico township officials claimed that growth decisions resided with the township water authority. When residents raised concerns about the effects of population growth on water supply, township supervisors suggested that development decisions were out of their hands, because it was up to the water authority to halt distribution of permits (Hernon 2003b, 2003d; Umble 2003). Divided governance provides an opportunity for blame shifting that can further impede coordination on complex problems.

Rohnert Park and Sonoma County: Consolidated Governance

Water's actually the biggest thing in land use right now.
—Rohnert Park City Council member

The Rohnert Park, California, case suggests that coordination problems might be more manageable if responsibility for land use and water is consolidated in a single government. A timeline for events in Rohnert Park appears in figure 6.6. Clashes over growth divided this Sonoma County city throughout the 1990s, but by 2000 the two sides in the "growth wars" had reached a compromise that would allow Rohnert Park to update its general plan. The compromise plan provided for the city to annex approximately 1,200 acres for new residential and commercial development, but it limited population growth to one percent annually for twenty years, translating into construction of approximately forty-five hundred new housing units. It also established a UGB around Rohnert Park that only a public vote could amend.[16]

In contrast to the divisiveness of growth issues prior to this time, the plan sparked remarkably little controversy within Rohnert Park. The plan's architects worked for two years to find a compromise between the higher growth rate sought by developers and the business community, on the one hand, and the lower rate that antigrowth activists supported, on the other. The final plan fully satisfied no one, but neither side mounted a serious challenge to it because it recognized the other side's power to block a competing proposal. A progrowth majority ruled the council at that time, but city residents had expressed their opposition to growth on earlier ballot measures. As a progrowth city official reflected in an interview, "Because the environmental community had been so successful with the electorate, if we didn't work for it, if we didn't approve it, we would have gotten nothing." An official on the other side who helped develop the compromise recalled that "some people were saying...that it would be better to have less growth. And the political reality was...there were potentially three votes against my perspective." Both sides were ready to accept the compromise in order to end the city's growth wars.

The compromise plan took into account Rohnert Park's resource constraints. The city council consulted with the water department all the way through the process, and the department gave regular reports at council meetings. The council expected to support the modest level of

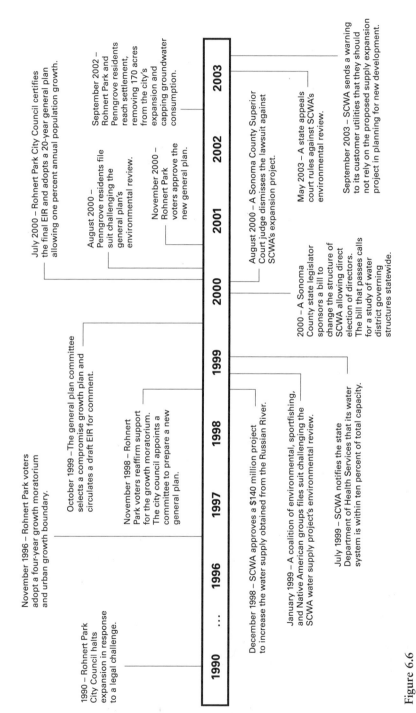

Figure 6.6
Timeline of Rohnert Park and Sonoma County events.

growth envisioned in the general plan through increased water purchases from the SCWA, recognizing that the city needed to lessen its reliance on groundwater pumped from municipal wells. SCWA had reached an agreement with its contracting communities that would substantially increase Rohnert Park's allotment by the time of the plan's buildout. The agreement allocated the new supply that SCWA expected to obtain from a planned water project on the Russian River, a project that had come under legal and regulatory challenge by the time Rohnert Park was reviewing its proposed general plan.

Although the challenges to the SCWA expansion introduced uncertainty into Rohnert Park's growth plan, the high level of coordination between the city council and the city's water department kept water issues off the agenda during consideration of the general plan amendment, and the plan enjoyed broad community support. The only opposition to the compromise proposal came from residents of Penngrove and Cotati, communities neighboring Rohnert Park. In written comments and public hearings, these neighbors raised concerns about the effect of Rohnert Park's expansion on water supplies, but they received no support from environmental and antigrowth groups within the city (see Callahan, 2000; Casey 2000; Sweeney 1999). After the city council approved the plan in July 2000, a group of Penngrove residents filed suit against Rohnert Park, alleging that the plan's environmental review relied on unproven assumptions about traffic and water impacts. Although the City of Cotati had raised concerns about Rohnert Park expansion during the planning process, its council voted unanimously not to join the lawsuit.

The Penngrove residents maintained that by counting on SCWA's planned expansion program, Rohnert Park failed to analyze the impacts of providing water for the growth envisioned in its general plan. Until the county completed its project on the Russian River, the city would have to increase pumping from municipal wells, depleting local groundwater resources shared by Penngrove residents. Should the SCWA expansion project never get built, common groundwater resources would be permanently dedicated to population growth in Rohnert Park. Calling the expansion project "paper water—not a sure thing but instead a hope," the Penngrove residents argued that "the magic wand of the SCWA is assumed to resolve all water problems" in the city's new general plan.[17] They also sought to restrict development on open-space areas that they claimed were important for groundwater recharge.

Rohnert Park responded that it had addressed these issues in its environmental review, noting in particular that it had adopted a policy mandating a groundwater-monitoring program and halting development if monitoring showed that further construction would produce a substantial decline in groundwater levels. The city argued that its neighbors were challenging the environmental review in order to pursue a larger goal: "In short, Petitioners would like to halt the growth contemplated in the City's General Plan."[18] Rohnert Park entered into settlement talks with the suit's petitioners, and the two sides ultimately reached an agreement in which the city abandoned a small portion of its annexation plan in the area closest to the Penngrove community and agreed to a cap on its groundwater consumption. The settlement allowed Rohnert Park to move forward with development.

In the three other cases where responsibility for land and water was divided between separate governments, reaching agreement on development disputes was costly. Dividing issue responsibility among independent governments creates institutional obstacles to informal cooperation and negotiation. General-purpose governments can pursue planning efforts without consulting relevant special districts; in these cases, by the time districts had the opportunity to comment, plans were already under public review. Instead of providing early feedback, special districts were forced to take formal positions against development proposals. Once negotiations became visible to constituents, officials in both general-purpose and specialized governments had an incentive to grandstand. They took inflexible positions and attempted to shift responsibility and blame for difficult decisions to the other venue. With public visibility, all officials became less likely to compromise, making it more difficult to resolve the dispute. As a result, water districts and municipal governments incurred transaction costs from bargaining over solutions, reconciling public and private positions, and pursuing litigation.

Rohnert Park largely avoided these costs by cooperating with its water department from the early stages of its planning process. The city faced a lawsuit on its land-use plan from residents of a neighboring community, but the case was weak because the neighbors could not appeal to a water supplier's publicly stated concerns as residents in Contra Costa County and East Cocalico Township could. Although municipalities engage in repeated interactions with the special districts whose jurisdictions overlap their own, they appear not to develop trust relationships with those districts that would help reduce transaction costs. Instead, it is the hierarchical relationship between municipal governments and their water departments that helps to minimize the costs entailed in settling these disputes and solving complex problems.

Consolidated governance of water and land use also curbed the participation of neighboring jurisdictions in Rohnert Park's planning process. When debate and negotiation between governments take place in public view, it is more likely that the conflict will expand to include neighboring jurisdictions as participants.[19] Local actors who oppose the land-use proposal may recruit neighboring governments to participate, or surrounding jurisdictions might act on their own out of concern about potential spillover effects of development on traffic, water resources, and quality of life. Neighbors become particularly engaged when they receive service from the same special district involved in the dispute. Consolidated governance offers a community's neighbors less chance to influence land-use decisions because planning and zoning proposals arrive on the public agenda fully formed. A Rohnert Park City Council member said in an interview that because the Penngrove residents arrived late in the process, Rohnert Park could not afford to start negotiating with them. And without service resistance from an outside water provider, the Penngrove petitioners were unable to convince the Cotati City Council to join their lawsuit. The absence of a public entity among the suit's petitioners reduced the perceived legitimacy of the suit's claims and reinforced perceptions that the complaints centered more on growth than on water supply.

By keeping much of the planning process outside of public view, consolidated governance reduces the level and scope of conflict over land-use outcomes. It lowers the costs of decision making and minimizes the opportunity for local actors to exploit water issues in order to pursue an antigrowth policy agenda. At the same time, a government that has authority over all local functions bears an especially heavy burden to be responsive and accountable to its constituents. Evidence presented in chapter 3 suggests that cities and counties sometimes fall short of this expectation. With many issues to occupy their attention, general-purpose officials might pay too little attention to the potential impacts of growth on the long-term sustainability of their water supply. Specialized governance provides an independent check on land-use officials' decisions, and it allows neighbors and interest groups to intervene and represent their interests.

In the disputes over development in Dougherty and Tassajara valleys, interest groups on both sides attempted to exploit the fragmented system of land use and water governance in order to achieve their goals. Developers took advantage of flexible special district boundaries to shop among multiple governments for service provision. Environmentalists sought to fight growth battles in water district venues because of the lower electoral campaign costs and favorable district boundaries that encompassed liberal constituencies. In Rohnert Park, governance of retail and wholesale water provision was consolidated, providing little opportunity for groups to select venues. Yet local actors still sought to exploit institutional design. When antigrowth interests in Sonoma County became frustrated with the county-run water supplier, perceiving the agency as biased toward development interests, they advocated for structural change to create a specialized water district.

While Penngrove residents were challenging Rohnert Park's general plan, environmental groups and their allies were pursuing litigation against SCWA's proposal to increase water diversions from the Russian River. The lawsuit focused on the diversion's impacts on fisheries, but in public statements the petitioners alleged that increasing water supply would provide a stimulus for more growth in the county.[20] SCWA had long drawn criticism for being secretive and unresponsive. The expansion proposal and perceptions that it was intended to fuel growth reinforced distrust of the agency within the environmental community, which began to call for a change in SCWA's governing structure. The agency is a dependent special district governed by the county board of supervisors. In effect, it operates like a county department. Charging that the supervisors' dual responsibility for water and land use constituted a conflict of interest, environmentalists and county antigrowth groups convinced a local state legislator to introduce a bill that would require SCWA directors to be elected at large.

As a result, at the same time that statewide environmental organizations were pursuing legislation to address the disconnect between water and land-use planning processes, Sonoma County groups sought to erect a boundary between the authorities governing water and land use. Critics charged that allowing the county board of supervisors to oversee the water agency put water in the service of the supervisors' progrowth policy agenda. They hoped that separately elected water district officials might be more favorable to their goal of blocking expansion of the county's water supply. As a county supervisor described, "I think there are a number of activists who see the infrastructure limits as being real limits and who are very nervous about relying on the judgments at any point in time of elected officials, and the way they're going to vote as being the factor that limits or they can count on to limit the growth potential in the area." Almost all of the county's growth had been occurring in cities, not in the county's unincorporated areas; thus, the question was whether the county water supplier would accommodate its contracting cities' land-use decisions. With the county supervisors' oversight, the agency pursued a policy that deferred to city planning processes. Sonoma County activists hoped that with an independent water supplier, they could pursue a strategy like the one their allies in the East Bay had used: challenging growth in a water district venue.

The effort to reform SCWA's structure was unsuccessful, but the groups continued to attribute SCWA's policy decisions to conflicts created by the agency's governing arrangements. Specialized governance of water in Sonoma County may have been somewhat more favorable to the interests of those who opposed growth, but it would have been unlikely to produce dramatic or sustained policy change. The results in chapter 4 showed that in high-growth regions, special districts rarely challenge the land-use authority of general-purpose governments. Even in the notable exception of the EBMUD case, the district's resistance to serving Dougherty and Tassajara valleys did not prevent valley development. As the Pennsylvania cases reveal, when water district and general-purpose boundaries coincide, governments have an incentive to cooperate. The governments serve the same constituency, and conflicts are less likely to expand outside jurisdictional boundaries.

Furthermore, antigrowth interests in Sonoma County would not necessarily have success electing their preferred candidates. Although in general it is less expensive to campaign for seats on functionally specialized local governing bodies, the SCWA proposal called for seats to be elected at large rather than by district, the same as for the county supervisors. The need to campaign countywide not only would be costly, it would also eliminate the opportunity to take advantage of district boundary design and compete for the most favorable seats. For all these reasons, reform in governing structure probably would have produced little change in the direction of decision making in Sonoma County.

Intergovernmental Coordination in a System of Specialized Governance

In recent decades, scholars of the new regionalism have called attention to the permeability of local political boundaries. Policies adopted by one city often have effects that spill over to its neighbors, and many of the most pressing problems cities face extend beyond their borders. Geographic boundaries that define local government jurisdictions can create policy problems and pose an obstacle to their solution.

Functionally specialized governance adds to these policy challenges. It creates crosscutting layers of jurisdictional boundaries, producing further externalities and requiring more stakeholders to address complex, large-scale problems. By increasing the number of local governments operating in a region and reducing the scope of authority for each, specialization calls for high levels of intergovernmental coordination in order to avoid the inefficiencies and political battles created by commons problems and negative spillovers.

This chapter's pessimistic message is that coordination can be difficult and costly. Sharing information across governments and engaging in joint planning require an investment of time and resources from all participants as well as a willingness to compromise. If local governments fail to make this investment at the early stages of a policy process, they end up negotiating over complex issues in the public sphere. Public negotiations make compromise more difficult by creating a political incentive for local officials to stand firm by their positions and shift blame elsewhere. In general, special districts seek to cooperate with cities and counties' land-use decisions, but when negotiations take place in public view, politicians come under pressure from constituents and interest groups to take inflexible positions. The results are increased barriers to coordination and the reinforcement of adversarial intergovernmental relationships that might already exist.

The chapter has an optimistic message as well, though. Broad participation and open decision making increase the time and cost of reaching decision, but they also help make policy processes more transparent. Moreover, specialized governance provides opportunities for appeal by local actors who have failed to achieve their goals in the original issue venue. Review of a policy or project in multiple venues may produce more creative and efficient solutions to complex problems. In many cases, the trade-off to having specialized governance is a more costly process for a result that has broader public support. Transaction costs become too burdensome sometimes, however, and a city or county can fail to coordinate with the relevant special district until a policy opportunity has passed.

Special district governance did not halt growth in northern California's East Bay, but it slowed down project approvals and substantially increased costs to developers—and consequently to families who bought homes in Dougherty and Tassajara valleys. These added costs led activists in Sonoma County to believe that specialized water governance would slow down growth in their region. By widening participation in the policymaking process, specialized governance might indeed help balance the local political power of prodevelopment interests.[21] On complex issues involving water and land use, however, delaying policy implementation will not always meet the preferences of those who favor slow growth and environmental protection. Time and resources spent on coordination, venue transfer, and legal battles may instead impede the progress of a water-conservation or watershed-restoration project. With fractured governance of multidimensional policy issues, resolution seems to come through threats rather than through cooperation, increasing the costs of agreement in the immediate case and potentially eroding trust for negotiation in the long term.

Even regarding situations where disputes have arisen, the cases presented here offer evidence that governments can reach agreement about how to proceed, and factors related to institutional design and problem severity may make it more or less difficult to find compromise. The shape of special district boundaries seems to have particular force in shaping interest-group strategies and the incentives for local officials. In Pennsylvania, water-authority boundaries coincide with township boundaries, so politicians governing land use and water serve the same constituency with the same distribution of policy preferences. Politicians may choose to ignore constituent preferences, as in the case of East Cocalico, but they do so at the risk of losing electoral support. Common boundaries prevent developers from shopping for services among multiple special districts, and they limit outside actors' participation when a dispute becomes public. In contrast, water district boundaries in the East Bay crosscut city and county boundaries, creating multiple constituencies with different preferences over water use and growth. Even purely responsive public officials therefore might disagree about the best course of action, and interest groups have an opportunity to influence outcomes through venue selection and strategic investment in lobbying and electoral campaigns. When cities or counties share a special district service provider, land-use decisions are more likely to have spillover effects on neighboring communities, widening participation in any potential dispute. In all of these ways, crosscutting boundaries make coordination more difficult by increasing the costs of reaching agreement.

In these cases, problem severity also affected specialized and general-purpose governments' behavior by changing public attentiveness to water issues and the stakes for policy inaction. In Pennsylvania, water has low public salience, but water-authority officials with limited functional responsibility nonetheless have an incentive to focus on water issues and deliver the policies their constituents prefer. Water authorities in both Pennsylvania communities discussed here were willing to challenge township officials in order to respond to constituent demands. In East Cocalico, the water authority would not jeopardize water supply to its existing customers by supporting a growth plan without assurance of an increased water supply, despite the township's firm commitment to pursuing new development. Adams Township Water Authority officials sought to expand the water system to provide service to existing township residents, rejecting township supervisors' plans for a developer-driven expansion. In both cases, conflict between the township and the authority attracted public attention and drew in some local residents. Water supply in Pennsylvania never approached the salience it has even during the wettest years in California, but temporary conditions of scarcity moved water issues to the top of the public agenda. When the conflicts began to threaten the acquisition of a long-term water supply and the township board's popular support, the potential benefits of reaching agreement helped overcome the costs of coordination.

In contrast, severe problem conditions and high water salience in the East Bay interfered with compromise between governments. This finding does not support my hypothesis that problem severity should make public officials more inclined to cooperate with one another. In the East Bay, drought conditions instead allowed local interest groups to exploit opportunities created by boundary design. By taking advantage of Alameda County voters' concerns about water resources, environmentalists won control of the EBMUD board and emboldened the district to challenge County Contra County on the Dougherty Valley development. Meanwhile, Contra Costa County continued pursuing its growth plan, a plan that county residents appeared to support. Neither government saw a potential benefit to cooperating until the salience of water issues in Alameda County waned.

Although public participation can hamper intergovernmental coordination, in many cases it is the only way that water availability gets considered in decision making about land use. Without a doubt, sustainable water management in the long term demands curbing growth in regions with limited water resources. It also requires allowing for growth where local resources can accommodate it. Throughout Pennsylvania and in many other regions, water supplies can support a growing population; the challenge lies in finding ways to distribute supplies fairly and efficiently. If activists succeed in blocking growth in communities where water supplies are plentiful, growth pressures mount in other communities where resources may be more scarce. Water availability must be a factor when planning for land use and growth, but the goal should be to direct growth to areas where local resources can sustain it.

The dominant approach to managing a local drinking water shortage was traditionally to increase water storage or to seek out a new supply source. Over time, imposing limits on new development gained popularity as an alternative solution. It is only recently that communities have started to give serious consideration to conservation and water reuse as viable strategies for stretching a water supply to serve more people. Land-use and water-conservation policies are not mutually exclusive—in fact, tightening the linkage between water and land-use planning can help to highlight opportunities for water conservation. Rohnert Park paid little attention to the water impacts of its rapid growth in the 1970s and 1980s. It was only when the city sought more water from the county wholesaler to accommodate the growth envisioned in its updated general plan that it agreed to retrofit homes with meters and adopt a rate structure designed to promote conservation. In Contra Costa County, environmentalists failed to block the extension of water service into Tassajara Valley, but the aggressive conservation requirements included in the final agreement became a model that groups are trying to replicate throughout the state. These cases are not isolated examples: in the first two years after implementation of California's water-adequacy law, one-third of all projects that came under review were required to introduce recycling or other conservation measures (Hanak 2005).

The spread of adequacy-screening requirements and growing attention to conservation may eventually help reduce intergovernmental conflict over land use and water. Even so, we can expect other issues to spark similar types of controversy. Local governments face social and environmental problems that are complex and cross issue boundaries. Fragmenting authority along geographic and functional lines makes it difficult to coordinate policy activity to address these complex challenges.

7

Specialization and Fragmentation in American Local Governance

A new local politics of water has taken shape in U.S. communities. Its cause is the growing scarcity of local water resources brought on by population growth, land-use change, and environmental limits to freshwater extraction. Although overall water use declined after 1980, depleted aquifers, environmental regulation protecting in-stream flows, and explosive growth in the nation's Sunbelt have heightened competition for freshwater resources. In contrast with other user groups that have reduced their consumption, the public drinking water sector has increased its water use during this period, leaving many utilities without enough water to quench their customers' thirst. Even public utilities with access to adequate supply may have trouble making use of it because of aging infrastructure for storage, treatment, and delivery.

Traditional supply-side solutions to local water scarcity have become less viable. Many surface and groundwater sources are already overcommitted, and the environmental and economic costs of new dams and reservoirs have ruled out countless projects that might have been feasible in an earlier era. Large-scale projects to increase water supplies have not disappeared altogether, however. A number of communities are experimenting with new technologies to develop water sources and expand storage capacity. Some are turning to the ocean or to previously unusable groundwater. More than two hundred desalination plants are operating in the United States; the largest plant is a $150 million facility designed to produce up to 10 percent of the water supply for the Tampa Bay region ("Applause" 2007). The San Antonio Water System has considered desalination of brackish groundwater from a neighboring aquifer as a way to reduce reliance on the overtaxed Edwards Aquifer discussed in chapters 5 and 6 (Needham 2007a). Other utilities are making use of treated wastewater. The Orange County Water District in southern

California invested $500 million in a new reclamation plant that converts treated sewage into drinking water (Weikel 2008). Aurora, Colorado, broke ground in 2007 on the $750 million Prairie Waters Project, which will send treated wastewater into the South Platte River, where it will be recaptured downstream and pumped back to the city for purification and reuse (Gertner 2007).

Such ambitious projects are not practical except for the largest public water utilities, which are few in number. Utilities serving more than one hundred thousand customers account for less than one percent of all community water systems. More than 80 percent of systems serve a population of thirty-three hundred or less, prohibiting the capital costs of large-scale supply projects (U.S. EPA 2002c). Some new supply technologies can be implemented at a smaller scale. Aquifer storage and recovery projects that involve injecting surface water or treated wastewater into groundwater aquifers for later retrieval allow more flexibility in supply management and avoid many of the costs of surface storage and transmission facilities. In addition, many communities are making use of reclaimed wastewater for landscape irrigation, either in specific applications such as a public golf course or through secondary water systems installed in new developments.

These supply-side solutions impose their own environmental costs. Desalination is energy intensive, and the reinjection of salty brine can be harmful to marine environments. "Toilet to tap" projects that allow purified wastewater back into the drinking water supply consistently attract public opposition based on concerns about their safety. Incorporating a river or aquifer into the water-recycling process—as in the case of the Prairie Waters Project and aquifer storage and recovery plans—may help to ameliorate public concerns, but fears of contamination persist (Dingfelder 2004). Aquifer storage and recovery projects can also disrupt seasonal stream flows and cause damage to riparian and wetland habitat.

Less costly and often more feasible are adaptive strategies that help localities live within resource limits. Critical to adaptation are demand-side policies, including economic incentives such as metering and pricing, regulations that ration water or limit new connections, procedures for integrating water and land-use planning, and education and outreach programs to promote water-use reduction through xeriscaping and appliance retrofit. In addition to encouraging (or mandating) lower water consumption, utilities can improve the productivity of existing resources by engaging in collaborative watershed management and developing cooperative agreements to allocate water more efficiently and equitably.

Attempting to adapt to resource limits rather than to overcome them represents an important shift in outlook for water managers, who tend to evaluate their own performance by their ability to meet consumer demand, regardless of its level, at any time of year. One California water manager characterized his job as making certain that limitations on water supply never impede additional suburban development (Lach, Ingram, and Rayner 2005, 2032). Adaptive strategies pose a fundamental challenge to these professional norms. The change in the water industry's guidance regarding progressive rates, as described in chapter 3, provides an example of the tension between industry values and contemporary challenges in water-resources management.

The water-management strategies outlined here require communities to make choices among multiple and competing values and uses for water. Communities must decide whether it is worth sacrificing green lawns in order to allow for continued growth—or, alternatively, whether they are willing to tolerate the housing shortage and economic slowdown that might follow a moratorium on new water connections. They must determine how to allocate costs between existing and incoming residents or among families with different lifestyles and patterns of water use. Communities must grapple with uncertainty about future population growth, water yields, and climate patterns. They also must decide what they owe to other communities in their region—if they are willing to scale back pumping in order to protect a shared groundwater aquifer or to give up water rights in order to distribute the burden of conservation more equally.

The new local politics of water is about how to allocate a scarce resource, and participants have complex and potentially competing goals. Environmentalists may aim for the lowest possible water use, even in regions with plentiful supplies of the renewable resource. They also might use water to obstruct development they oppose on broader grounds. Water managers may want to capture efficiencies to protect the financial health of a water system or to meet customer demands for a cheap, unlimited supply. They also may promote conservation as a means to allow system expansion, which creates a problem if other aspects of the local environment cannot sustain a larger population (Martin, Ingram, Laney, et al. 1984). Residents typically want unlimited water at a low price, but they are increasingly recognizing the existence of resource limits, especially in locations where those limits are closer in sight. Public officials typically are risk averse, hoping to satisfy all constituencies. Indeed, a policy solution sometimes does exist that can satisfy multiple, competing goals. At other times, local leaders must make difficult choices and elevate some interests over others.

In the new local politics of water, public utilities play a critical role. Resource scarcity and environmental regulation have brought water utilities into policy territory that previously lay outside their domain. They must interact with planners, community groups, the press, and neighboring jurisdictions. They must be attentive and responsive to diverse points of view. This book has examined whether the growing specialization and fragmentation in local governance interferes with the policy coordination and responsiveness called for by current conditions. I have used the case of drinking water policy to argue that in order to understand the consequences of specialization, we must consider the status of policy problems and specific institutional structures that promote or inhibit government accountability. The next section reviews and elaborates on these findings, and then I consider their implications for local governments' capacity to promote sustainable development and confront global climate change.

The Effects of Specialized Governance

The previous four chapters examined specific controversies and public policies to assess the impact of specialized governance on the management of drinking water. A consistent theme has been the contingency of specialization's effects. Table 7.1 summarizes the findings reported in these chapters. As the table demonstrates, the findings lend strong support for a conditional theory of specialized governance. Separating one government function from the other activities that cities and counties usually manage does make a difference for policy outcomes, but its effect varies across local contexts. On the whole, I have found that special districts are more responsive to their constituents than the conventional wisdom suggests, but that does not make them a simple fix for local public problems. Policy questions that cross issue boundaries pose a challenge for specialized governance, and fragmentation of authority introduces new actors into the policy process who represent multiple political constituencies. Interaction among governments and groups in an institutionally fragmented system may produce outcomes unanticipated by those who originally backed a special district's formation.

Table 7.1
The Conditional Theory of Specialized Governance: Summary of Findings

Conditioning variables	Predicted effects	Findings
Problem severity	*Responsiveness:* ↑ responsiveness of cities and counties ↓ policymaking by special districts outside functional boundaries *Intergovernmental coordination:* ↑ benefits of cooperation	*Support:* chapters 3 and 4 *No support:* chapter 5 *Mixed results:* chapter 6
Special district elections	*Responsiveness:* ↑ responsiveness of special districts *Intergovernmental coordination:* ↑ scope of conflict on complex issues ↑ benefits and costs of cooperation	*Support:* chapters 3 and 6 *Support:* chapters 5 and 6
Special district boundary flexibility	*Intergovernmental coordination:* ↓ benefits of cooperation	*Support:* chapter 5
Contiguity between special district and city or county boundaries	*Intergovernmental coordination:* ↓ scope of conflict on complex issues ↓ costs and ↑ benefits of cooperation	*Support:* chapter 6

Conditioning Effect of Problem Status

The chief criticism of special districts is that they serve the private interests of those most active in their formation—typically, the local development community—and consequently help a progrowth minority achieve its policy goals. Critics charge more generally that special districts' low public visibility produces a bias that favors those actors who have the time, money, and political experience to monitor district activities. If these assessments are accurate, water districts should be unwilling to adopt new management strategies opposed by developers and wealthy homeowners. Policies that are unpopular with these groups include rate structures that shift the burden of costs onto customers with the heaviest use, development fees that make growth pay its own way, and processes that weigh water considerations more heavily in decisions about land-use policy.

My analyses, however, show that specialized governance does not necessarily serve the interests of progrowth minorities. In fact, where the stakes are low and problem status is not severe, actors pursuing progrowth outcomes have more success in general-purpose venues. A city or county council's crowded agenda allows developers to dominate outcomes if water is plentiful and growth rates are low. Dedicating a venue to a single function frees policymakers from competing demands on their attention and allows them to be more responsive to public preferences. Where governments must grapple with a serious problem, though, specialization does not have the same effect. More serious problem conditions draw city and county officials' attention and align their incentives with those of specialized policymakers. In chapter 3, I showed that climatic conditions, which are a critical determinant of local water scarcity, have a differential impact on how generalist and specialized officials balance the provision of public goods against the imposition of private costs when making rate-setting decisions. City or county utilities located in hot climates are more likely to serve majority interests and provide public benefits than are their counterparts located where climate conditions are moderate. Problem severity elevates an issue over other local functions on a crowded city or county agenda, but it has little effect on the decisions of special water districts, which always have an incentive to respond to the majority.

Chapter 4 analyzed a policy choice that crosses issue boundaries through its impacts on both a water system's revenue structure and a community's approach to land use. When a water system imposes high impact fees, it helps shape a community's land-use policy by increasing the costs of development and sending the message that growth should pay its own way. I demonstrated that the pressure of a high growth rate reduces the impact of water governance on fee levels. Where growth rates are low, specialized governance increases reliance on fees that require developers and incoming residents to fund the costs of infrastructure expansion. By intensifying conflict over local development policies, rapid growth lessens the influence of governing structure. Developers and sometimes their antigrowth opponents monitor and attempt to influence all policy decisions that have implications for land use. Cities and counties become more responsive to constituents who call for relief from paying the costs of growth, and special districts respond by deferring to general-purpose governments' land-use authority. Surprisingly, the analysis of cooperative agreements in chapter 5 lent no support for the hypothesis that severity would promote cooperation by increasing its potential benefits. This negative result suggests that local officials are responding to the configuration of political interests that emerge under severe problem conditions rather than to problem severity itself.

Problem status has an impact on local officials' behavior because of the enduring possibility that constituents will hold officials accountable for problems that go unsolved. Politicians perceive that a more severe problem is more likely to become salient for the public, and they worry that they will suffer electoral or reputational damage if they fail to address the problem. This assumption is reasonable in the water case. Americans expect to have access to safe, plentiful drinking water at a low cost, so they give little thought to water when it meets their expectations. Where resource scarcity jeopardizes the quality, quantity, or cost of water, however, the public will pay attention and call on politicians to develop policy solutions.

Cross-sectional comparison within large national samples captures the wide variation that exists in problem conditions. The samples include communities such as those in the southern California high desert with extreme climate conditions and booming rates of growth, places in the Rust Belt where water is plentiful and growth rates are low, as well as many communities that fall somewhere between these extremes. But problem conditions also can vary over time within a single community, regardless of whether overall conditions are severe or relatively secure. Aging infrastructure, vulnerabilities in the supply source, weather events, and even modest changes in local demand can create a water supply problem where none existed previously. Water supply, like other local issues, moves on and off the public agenda, capturing public attention and then losing it, to varying degrees according to the baseline level of public interest and problem severity. Objective conditions are not the only factor influencing variation in the salience of local issues over time. Nonetheless, a severe problem offers a compelling message for interest-group mobilization and raises the stakes for the politicians who are designing policy solutions.

The cases in chapter 6 revealed some of the mechanisms by which change in problem severity over time produces change in issue salience. Drought in California and local water shortages in Pennsylvania drew public attention to the limits on water resources and the relationship between water and growth. In Pennsylvania, water salience created an incentive for public officials to overcome their policy dispute and reach a compromise on future growth plans.

Problem status affects interest-group activity as well, because severe problems can attract new actors into local politics or stimulate more group participation. In California's East Bay, issue salience expanded the scope of conflict enough to interfere with intergovernmental coordination. It was only once the drought ended and the broad public turned its attention to other issues that the governments involved could settle on a policy solution. Similarly, booming growth can pose an obstacle to coordination by increasing the prevalence of developers and landowners in local politics and providing more incentive for community members to overcome collective action problems and organize to slow the pace of development.

The assumption that problem severity produces more potential salience should hold for many of the functions that special districts perform. These functions tend to have widespread and immediate impact on the lives of community residents, so people can perceive when conditions are worsening. Moreover, special district functions often have low ideological content and spark little disagreement over whether and how government should be involved. For functions such as flood and fire protection, mosquito abatement, and public transit, the public expects effective government performance. Where natural or social conditions exacerbate the policy challenge, the issue receives more public attention. The empirical tests here have focused on exogenous determinants of problem status, but the BexarMet case shows that governance itself can also affect problem severity through failure—or success—in addressing a policy challenge. My results provide no reason to believe that policy failure is any more or less common among special districts than among general-purpose governments. The relationship between problem status and salience might not hold for problems that affect a minority of the population, such as homelessness, or for issues that elicit differences of opinion about government's proper role, such as economic development. For most local functions that allocate goods and services among community residents, however, I would expect the severity of a public problem to condition the impact of special district governance on policy outcomes.

Conditioning Effect of Institutional Design

The findings described in the previous section confirm predictions from public choice theory that low-profile issues will get lost in a general-purpose policy venue, creating opportunity for interested stakeholders to dominate decision making. In regions where water is plentiful, utilities operated by cities and counties are more likely to deliver outcomes that favor development interests. The policies that water districts adopt under these conditions are more consistent with majority preferences.

At the same time, my results lend support to concerns raised by metropolitan reformers about political invisibility. Many special districts have an institutional design that interferes with policy responsiveness. Although specialization tends to reduce bias in outcomes where problem conditions are normal, I demonstrated in chapter 3 that this effect is less pronounced for districts whose officials are appointed rather than elected. Moreover, results in chapter 5 indicate that elections can promote interlocal cooperation by providing an incentive for politicians to seek out assistance in addressing local service gaps. Even though special district elections attract very low rates of participation, they provide a mechanism for citizens to influence a district's policy direction. During the California drought in the early 1990s, water issues became salient enough to East Bay residents that many of them turned out to cast ballots for the EBMUD and Zone 7 governing boards, sending a clear message in support of a more conservation-oriented policy. On the whole, I find that without direct elections to provide accountability, local officials expend less effort to provide public goods and solve policy problems.

Boundary design also conditions the impact of specialization on policy-making. In chapter 5, I showed how the stringency of requirements for boundary expansion affects the relative attractiveness of different options for addressing gaps in water service. Flexible boundaries create incentives for special districts to annex new territory and enlarge their own jurisdiction rather than develop cooperative agreements with neighboring governments for service provision. Interest groups also attempt to take advantage of boundary design and flexibility. Chapter 6 explored in detail how both sides in a growth dispute attempt to exploit the intersecting boundaries of overlapping special districts in order to achieve their policy goals. Specialized governance of essential local services increases the number of veto points for a development proposal, but it also allows project developers to shop for service providers. Where special district boundaries crosscut those of general-purpose governments, there is more opportunity for strategic venue choice, and it is easier for neighboring governments and residents to get involved in a local dispute. Intergovernmental coordination becomes more difficult, and politicians have an incentive not to compromise. Outcomes are ultimately contingent on the location of district boundaries and the geographic distribution of public attitudes about growth.

Implications

The conditional nature of specialization's effects has a number of implications for how we evaluate the costs and benefits of special district governance. First, it demonstrates the critical point that special districts are not all the same. The fiscal and administrative autonomy that distinguishes an independent district from a dependent district reduces accountability to the actors who created the district but also has the potential to heighten public accountability. We need to be attentive to the difference between districts that are subordinate to another public entity and independent governments with their own budgets, facilities, and governing officials. The specific institutional structures of an independent special district also have an impact on its approach to public problems. Although independent districts have the potential for accountability and responsiveness to their constituents, they best realize this potential when accountability mechanisms are in place and the boundaries of their authority are clear. Elected special district officials are more likely than their counterparts on appointed boards to engage in interlocal cooperation and to deliver policies that are congruent with majority preferences. Even if special district elections attract little public participation, the possibility of electoral punishment influences district officials' behavior. The boundaries of responsibility also make a difference for special district performance because interest groups will attempt to exploit any ambiguity in functional or geographic scope in order to pursue their policy goals. A district with clearly defined responsibilities can assert its independence and authority when taking action on policy questions.

Given the diversity among special districts in their functional activities, revenue sources, and institutional structure, the statement that "special districts are not all the same" should come as no surprise. Indeed, Kathryn Foster concluded a decade ago in her important book on special districts that "'special districts' is a plural" (1997, 218). Yet theoretical approaches to specialized governance have not accounted for this diversity. They assume either the presence or absence of democratic procedures and boundary flexibility, when in fact those features are variable across special districts and play an important role in shaping outcomes. Specialization is most likely to produce public benefits if electoral procedures are in place to provide direct public accountability.

Second, these results shed light on how local institutions mediate contemporary political conflicts over growth and land use. Much of the theorizing about special districts assumes a model of local politics in which developers, landowners, and other economic elites dominate decision making about development and land use (Logan and Molotch 1987). This model depicting a unified growth elite against the diffuse majority no longer applies in many communities that have seen the emergence of vigorous antigrowth movements. Residential associations and community and environmental groups have become more active in organizing against continued development, supported by rising public dissatisfaction with growth's environmental costs and its local impacts on quality of life. Opposition to the growth machine now has its own political force in many communities and its own ability to monitor and influence a less visible government. The two sides in a growth dispute have different bases of support. A progrowth coalition typically is united by an economic interest in land development; an antigrowth coalition may comprise both neighborhood activists with a geographic foundation for their opposition and environmentalists interested in broader public goods (Gerber and Phillips 2003; Lubell, Feiock, and Ramirez 2005). Different elements of an antigrowth coalition may or may not agree about the merits of specific project proposals.

The growing political force of antigrowth movements requires that we loosen assumptions about the configuration of interests involved in development conflicts. Both supporters and opponents of a proposed project might sponsor organized advocacy efforts, and they will take advantage of the boundaries and design of governing institutions in order to find the venue most favorable to their policy goals. Environmental and community groups also might consider taking growth proposals directly to voters, a possibility that encourages local officials to take positions more closely reflecting constituent opinion (Gerber 1996; Gerber and Phillips 2003, 2005). Along with problem severity and institutional design, the configuration of local interests conditions the impact of specialization on policy outcomes. Under the conventional wisdom that special districts exhibit a progrowth bias, environmentalists and community groups' efforts to shift disputes into special district venues and even to create a new special district, as reported in chapter 6, would seem self-defeating. The conditional theory makes sense of these efforts by showing that organized antigrowth groups may prefer a specialized to a general-purpose venue if the boundaries define a constituency that is more supportive of antigrowth goals. Furthermore, where growth rates are low and resources plentiful, policymaking related to land use and urban services can become dominated by development interests in general-purpose governments while local officials dedicate their attention to more salient issues. Specialized governance, in contrast, may allow an unorganized majority to overcome the advocacy efforts of those who favor growth.

The third implication of these findings is that special districts are not to blame for most failures of responsiveness in local policymaking, nor are they an easy fix that will solve most public problems. The effects of specialized governance are complex and contingent, and an autonomous special district will not necessarily serve the interests of the actors who created it. Developers support special district formation as a means to obtain public services without incurring private financial risk. They also might anticipate that districts' political invisibility and bonding powers will make it a more development-friendly venue in the long term. My results provide no evidence that districts consistently support a progrowth agenda, however. Conversely, despite widely held perceptions that special districts favor developers, environmentalists recently have attempted to pursue their policy goals in special district venues. The conditional theory indicates that only within certain contexts will specialized governance offer benefits for these groups, and the benefits may be balanced by gains for opposing interests. Some large special districts get formed as a means to overcome coordination problems and regionalize service delivery. My findings suggest that they may instead make policy coordination more difficult.

Special districts can help communities provide a service at a level or scale that might not be feasible otherwise, especially in the face of state-imposed legal restrictions on local government organization or local fiscal flexibility. Special district formation is often the best option for meeting changing service demands or for addressing an emerging public problem. Even so, specialization does not guarantee better or worse governance, and, as noted, a special district may not serve its creators' interests. The contingencies of problem severity and institutional design make it difficult to predict in general terms how specialization will affect policymaking in the long term. However, they do provide guidance for anticipating the impacts of special district governance within a specific set of institutional and problem conditions. Moreover, for problems that most need addressing, governing structure is likely to have little impact at all.

A fourth implication is that states can do more to improve the performance of local governance systems. Calls for local government reform are often too sweeping, demanding large-scale consolidation within or across functional boundaries. These efforts are rarely successful, which may be for the better because they propose universal reforms to governance problems that are conditional in nature. In many contexts, special districts perform well, and by some measures they may be better than an alternative structure. Nonetheless, because states determine the rules governing local government operations, they can provide incentives for greater responsiveness and intergovernmental coordination. Most important, requiring direct election for special district boards would increase accountability and encourage district officials to govern more effectively, even if participation rates in special district elections are low. States also might be more attentive to the importance of boundary design. In some cases, the efficiencies achieved from crosscutting jurisdictional boundaries outweigh any coordination costs. If a district's function has a clearly defined geographic scope, or if a particularly high value is attached to public participation in policymaking, then crosscutting boundaries may be the preferable design. In other contexts, however, states might want to align special districts' boundaries with general-purpose jurisdictions' boundaries in order to create common constituencies and to promote cooperative relationships. Restricting a special district's ability to expand is another way to promote cooperation where resources are distributed unevenly across communities.

Apart from these institutional reforms, states can set up processes requiring coordination on specific policy issues. States have an important role in setting the institutional framework for local water management (Blomquist, Schlager, and Heikkila 2004). We saw in chapter 6 that state water-adequacy requirements have compelled localities to overcome coordination costs and to integrate water and land-use planning processes. Similar requirements might promote cooperation in transportation and land use, for example, and depending on the local appetite for participation, states might create more or less restrictive participation requirements in environmental review laws. As local governments mediate local battles over development proposals, states have the opportunity to direct overall growth to the regions best able to support a larger population. Either by mandate or by incentive, a state with well-defined goals for growth patterns and local government performance has the tools to promote its vision.

Finally, these results suggest that our evaluation of specialized governance ultimately depends on the values we seek to maximize. Under many circumstances, special districts offer a trade-off. They provide more opportunity for public deliberation but may hinder the development of cooperative policy solutions. They can be more responsive to majority preferences, but this responsiveness may interfere with equity and sustainability in policy outcomes. Responsiveness to a single constituency's preferences can have negative consequences for a larger population. This outcome holds true particularly in metropolitan regions, where local governments' responsiveness to constituents' demands for higher property values, lower taxes, and more homogeneous communities has contributed to racial segregation and socioeconomic inequality across jurisdictions (Altshuler, Morrill, Wolman, et al. 1999). Majority rule can produce policies that impose the burden of costs on those who are not represented, as chapter 4 described in the case of development impact fees. It provides little protection for minority rights or for the interests of those who live outside the jurisdiction.

The value of responsiveness is no less relevant for water than for other public functions. Water has important social and cultural meaning. It might define territory, provide sustenance and security, demonstrate power, create wealth, or offer spiritual or recreational benefits (Blatter and Ingram 2001; Ingram 1990). Governing structures for water must be accessible and responsive to the public, so that local officials can understand and appreciate the multiple and potentially competing demands on local water resources. But responsiveness is not the only criterion we should use in evaluating systems of governance, especially if we find that responsiveness threatens other shared values. Drinking water is an essential good. Policymakers must protect universal access to water, even if it means overriding a majority preference for unlimited consumption. Water officials also need to consider how current policies will affect resource availability in the long term. A local majority rarely will support conservation policies that limit the quantity of water residents can consume, but usage restrictions may be necessary to safeguard a community's water supply. These tensions extend to other functions that special districts perform. A government seeking only to satisfy a majority might cancel transit routes for poor communities that need them the most or fail to fund levee improvements for neighborhoods most vulnerable to flood. Local actors engaged in institutional design should begin by defining their goals for public-service provision. The marginal gains in responsiveness that specialized governance offers on low-salience issues may not be worth potential losses in equity and intergovernmental cooperation.

As populations grow and policy problems become increasingly complex, special districts are a logical governance response. They allow decision-making institutions to be crafted to the needs of a specific policy problem. For issues that would not capture the attention of busy city or county officials, a special district structure allows a tighter connection between public preferences and the government's policy decisions, helping to overcome the bias that typically exists in local policymaking. As special districts proliferate, however, what the relevant public is for a policy decision becomes less certain. Notions of political community are more fluid, interest groups make strategic use of venues, and policy questions that cross functional lines pose a significant challenge to governing structures. The lesson of this study is that the impacts of special district governance are neither simple nor universal. The more we understand the political context for specialization, the better we can predict its effects.

Special Districts, Sustainability, and Global Climate Change

The new local politics of water is a product of the transition to sustainability as a guiding goal in the management of public drinking water. Where water systems previously sought to satisfy customer demand above all else, they are increasingly balancing that goal with attempts to lessen the impacts of drinking water provision on the ecological health of natural resources. Sustainability measures the degree to which activities maintain the earth's—and the local region's—carrying capacity rather than depleting it. Sustainability policies aim to promote economic growth without causing environmental damage or undermining quality of life for current or future generations. In the case of water, a key indicator is whether resources are being depleted faster than they can be replenished. Managing drinking water sustainably also requires financial investment to protect water quality and delivery systems for the long term.

The findings presented in this book demonstrate that special districts are able and willing to pursue policies that promote sustainable water management. The question remains, however, how specialized governance affects broader community efforts to achieve sustainability. Over the past two decades, sustainability has emerged as an organizing principle for national efforts to protect the environment and for local economic and community development (Mazmanian and Kraft 1999; Portney 2003). Although the fundamental concept of sustainability is global in scale, it can be achieved only through local action. Agenda 21, the international plan for action on sustainable development, emphasizes the importance of local initiatives:

Local authorities construct, operate and maintain economic, social and environmental infrastructure, oversee planning processes, establish local environmental policies and regulations, and assist in implementing national and subnational environmental policies. As the level of governance closest to the people, they play a vital role in educating, mobilizing and responding to the public to promote sustainable development. (United Nations Conference on Environment and Development 1992, 28.1)

Dozens of large cities throughout the United States have launched formal efforts to reduce their environmental impacts and to live within local resource limits (Portney 2003). These efforts are ambitious in scope, endeavoring to address water consumption and pollution along with a host of other impacts stemming from a community's transportation systems, waste management, land use, food supply, and building maintenance. Many local efforts define sustainability to encompass broader goals as well, such as environmental equity, social justice, and economic prosperity.

To achieve sustainability requires new ways of confronting problems and designing public policies. Sustainability initiatives involve multifaceted consideration of the environmental, economic, and social aspects of a problem. They acknowledge the interdependence among issues and attempt to integrate policy efforts across functions and levels of government. In a sense, the sustainable cities movement is about making functional boundaries more porous. It aims to lessen a transportation project's impacts on groundwater replenishment, for example, or to ensure that port improvements do not bring about plant siting decisions that result in environmental injustice. Accounting for these complex interrelationships poses a significant challenge for governing systems that are organized along functional lines.

Sustainability efforts rely on collaboration and cooperation among governments and other actors. Even governments with a broad functional scope have limits on boundaries, expertise, and authority that would prevent them from achieving sustainability goals on their own. Sustainability policies must be flexible and adaptive to changing conditions, and they will be effective only with participation from a broad group of public- and private-sector actors. Most cities that have introduced sustainability initiatives attempt to build consensus among community residents and organizations about goals and strategies. Where governance is fragmented among multiple special districts, participatory processes are even more important in order to secure the commitment of the independent authorities that manage the community's infrastructure and natural resources. These processes are costly and time-consuming, however, so special districts must be convinced that the potential benefits of sustainability planning warrant the effort.

A broad community commitment to sustainability does not come without costs. Sustainable development calls for policies that require individual sacrifice in order to provide public goods, and it may entail overcoming local opposition to development in order to direct growth to areas where resources can support it. Residents and businesses have little reason to back initiatives that do not serve their self-interest, and local officials have more incentive to respond to constituent opinion than to cooperate with neighboring jurisdictions on abstract policy goals. Consensus-building processes may help residents to see the benefits of policies that promote conservation and social equity. Nonetheless, it is difficult to conquer the motivation to overuse common pool resources and ignore externalities imposed on neighbors, especially in a region with complex, overlapping jurisdictional boundaries. Sustainability can be achieved only with local authorities' participation, but state or federal involvement may be necessary to induce local governments to participate. Through a combination of incentives and mandates, higher levels of government can change the balance of costs and benefits involved in cooperation and can encourage communities to think more broadly about community development and environmental protection.

A key element of sustainability planning involves local efforts to address global climate change. These efforts include mitigation programs intended to reduce greenhouse-gas emissions as well as strategies to adapt to expected climate change impacts. For most local functions, special districts can do a great deal on their own to achieve emissions reductions. They can improve the energy efficiency of their buildings and equipment, invest in green power and fuel-efficient vehicle fleets, and provide incentives for employees to reduce waste and use public transportation. Some special district functions provide opportunity for more significant gains: water districts can increase the efficiency of their treatment systems; special districts operating electric utilities can increase their reliance on renewable energy sources; public housing districts can conform with green building standards; transit districts can invest in green bus fleets and implement strategies to increase ridership; and waste-management districts can install systems to capture and reuse landfill methane gas. In these and a variety of other ways, local governments can act independently to help protect the global climate. Both the potential for financial savings from efficiency improvements and an awareness of direct threats that climate change may pose to a community can help build public support for climate-protection policies.

Local efforts to mitigate climate change do not necessarily require policy coordination among governments. In contrast, planning for adaptation demands a broad view of potential impacts and a collaborative approach to address new risks and realities. The management of drinking water has traditionally assumed that a region's hydrology is stationary and then designed for changes that might occur in land use and patterns of demand. The assumption of stationarity is no longer valid, however. Climate change introduces the possibility of new water quality problems, changed stream flows and groundwater recharge rates, and more frequent and more severe droughts and flood events (Kundzewicz, Mata, Arnell, et al. 2007). According to the Intergovernmental Panel on Climate Change, water resources in the United States have already been affected by climate change (Field, Mortsch, Brklacich, et al. 2007). Annual evapotranspiration rates have increased nationwide. Stream flow has increased in the eastern United States over the past sixty years, but declined in the western mountains. Overextended freshwater resources in the Colorado and Columbia rivers and the Ogallala Aquifer are especially vulnerable to changing climate conditions, and lower water levels in the Great Lakes are also expected to have impacts, including water quality problems and increased competition over freshwater supply. Adaptation to these changing conditions will demand integrating management of natural resources across geographic and institutional boundaries. Policymakers will need to account for the relationships between water quality and quantity, as well as between management of land and management of water. Coping with the hydrologic effects of climate change will require coordination among the authorities responsible for land use, flood control, watershed management, wastewater treatment, and drinking water supply. In many communities, this coordination means cooperation among multiple, autonomous governments.

The prospect of global climate change highlights the importance of problem severity and context when evaluating governance systems. The policy challenges that communities face will depend on local conditions; they may include adapting to changed timing and level of stream flow, reducing reliance on depleted groundwater reserves, negotiating major water transfers across distance and between uses, or controlling outbreaks of waterborne disease brought on by extreme precipitation. Water issues are likely to become more severe in communities already struggling with water scarcity, but climate change may also introduce water challenges in communities that do not expect it. For example, one study predicts that the New York region's smaller water systems may become vulnerable to greater supply variability (Field, Mortsch, Brklacich, et al. 2007). We have seen that the incentives for cities and counties to grapple with water issues and respond to public demands vary with the severity of water scarcity. It is possible that water problems brought on by climate change will catch some local systems unprepared, and these systems may thus fail to address the changing conditions of water availability until it is too late.

High levels of uncertainty surround assessments of the possible impacts of climate change on water resources. Although early action is essential if we are to meet these challenges, the character of risk might change over time. Policy approaches must emphasize flexibility and adaptability to changing problem conditions. The uncertainty and the potentially broad scope of impacts call for participatory local processes to consider potential scenarios and opportunities for policy coordination. As with local sustainability initiatives, however, these processes can be costly and time-consuming, even more so in a context of specialized and fragmented governance. Considering the high stakes of policy inaction, it is critical that local governments start preparing now.

Appendix 1
Explanation of Data and Model, Chapter 3

Data

Data on rate use and many of the models' explanatory variables come from the 1999 Financial and Revenue Survey conducted by the AWWA (1999). Omitted from the analysis are privately owned water systems, utilities operated by a state or federal government, ancillary water systems, and utilities that only sell water wholesale. In addition, only utilities that report using uniform, increasing block, or declining block rate systems for residential users are included, leaving out water systems that do not meter water use and those that employ only seasonal or peak rates.[1] After the sample is narrowed with these omissions and missing data in some cases, the remaining sample for the main model includes 427 utilities: 322 operated by cities or counties and 105 by special districts.

The AWWA survey includes a question asking whether utility governance is specialized or general purpose. I confirmed utility responses using local government Web sites, U.S. Census Bureau data, and review of state laws defining a water district's scope and responsibilities. I calculated control variables on the water system's supply source, financial health, and customer base directly from the survey data. The operating ratio is a standard accounting measure calculated by dividing total operating revenues by operations and maintenance expenses. It appraises whether a utility's operating revenues are sufficient to cover its costs. As a general rule, a higher operating ratio suggests better financial health, although there are reasons why a financially sound utility might operate at a low ratio in a given year. Utility financial data were the largest source of missing data from the AWWA survey. Removing the operating ratio from the model increased the sample size to 452 utilities and did not change the substantive results.

Climate data come from maps produced by the National Climatic Data Center (NCDC) showing annual mean total precipitation and mean daily maximum temperature, computed for the period from 1961 to 1990. The NCDC integrates point measurements collected at thousands of weather stations nationwide with other spatial datasets to generate indices of climatic elements and to map their spatial distribution. Using data from the Census Bureau's 2000 U.S. Gazetteer, I geocoded the centroid of the city, county, or county subdivision in which each utility is located and plotted the utilities as point data on a map of the United States. Merging the climate maps with point data on utility location produced index values on precipitation and daily maximum temperature for the utilities.

Data on procedures for selecting local government officials come from the Government Organization File of the 1987 *Census of Governments* (U.S. Census Bureau 1990). The same source also provides the information on city and county form of government that is included in the ward elections model for general-purpose governments. The variable *council/manager system* includes both city council/manager and county council/administrator systems, and *executive/council system* combines city mayor/council systems with county council/elected executive systems. The base category includes commission and town meeting forms of government. Probit estimates of the effects of election rules appear in table A1.1.

Model

The main text presents results from a probit model that assumes the AWWA sample was randomly selected from the universe of public retail water utilities. In fact, the utility survey was designed to describe the AWWA membership and thus does not constitute a representative sample.[2] The AWWA sent questionnaires to approximately 3,400 of its 4,400 members, and it tended to drop its smallest member utilities from the sample. Response rate for the survey was 21 percent. As a consequence of this selection process, the AWWA sample is skewed toward the largest utilities as compared to estimates of the population of retail water systems (U.S. EPA 1997a). The sample also overrepresents utilities that rely on surface sources.

This appendix presents a Heckman probit model to account for the AWWA survey's nonrandom sample-selection process and to test

Table A1.1
Probit Estimates: Effect of Election Rules on Adoption of Increasing Block Rates

Variable	Election of special district board members	Ward elections: Cities and counties	Ward elections: Special districts
Temperature	.55**	.13	−.37**
	(.27)	(.08)	(.15)
Elected board	2.72**		
	(1.11)		
Elected board * Temperature	−.68**		
	(.29)		
Proportion elected by ward		−1.57***	−3.13**
		(.54)	(1.42)
Proportion ward * Temperature		.37***	.70**
		(.13)	(.32)
Precipitation	−.10	.03	−.23
	(.13)	(.09)	(.15)
Operating ratio	−.00	.18	−.28
	(.18)	(.15)	(.34)
Surface water	−.49	.17	−.61*
	(.38)	(.20)	(.47)
Purchased water	−.22	.22	−.34
	(.46)	(.25)	(.54)
Proportion retail sales	.71	1.76***	−.08
	(.78)	(.63)	(1.01)
Urban	−.04	.29	−.29
	(.33)	(.21)	(.45)
West	.14	.75**	.33
	(.42)	(.36)	(.47)
Midwest	.01	.10	
	(.67)	(.29)	
South	−.35	−.37	1.49*
	(.59)	(.35)	(.97)
Council-manager system		−.17	
		(.28)	
Executive-council system		−.58**	
		(.29)	
Constant	−1.95	−3.15***	3.49*
	(1.41)	(.96)	(1.82)
N	95	318	66
Pseudo R^2	.12	.22	.15

Notes: Robust standard errors in parentheses. *$p < .10$, **$p < .05$, ***$p < .01$ (two-tailed).

the robustness of the results shown in the main text. The Heckman model addresses the possibility that the factors underlying selection of a nonrandom sample also influence the dependent variable in an analysis. If this is the case, then ignoring the selection process is equivalent to omitting a variable and produces inconsistent estimates of parameters (Achen 1986; Dubin and Rivers 1989; Greene 2003). The model requires data on the full universe of water utilities in order to predict selection into the AWWA sample. These data come from the Public Water System Inventory, an EPA-maintained national inventory of public water systems with information about population served, source water, location, and ownership. The universe was trimmed to include only active water systems that are located in a state (not on tribal land or in a U.S. territory) and owned by a local government (not privately owned or owned by a state, federal, or tribal government), leaving 22,118 observations to merge with the AWWA data. The log of population served by the utility operates as the exclusion restriction, a variable that influences selection but not the dependent variable of interest. It is an appropriate instrument because it has a powerful effect on selection, but no predicted relationship with rate adoption.[3]

Results from the Heckman model appear in table A1.2, along with estimates from the probit model that produced the marginal effects shown in the main text. When the results from the two models are compared, it is evident that the estimates are not conditional on modeling the sample selection process. Coefficients in the outcome equation of the Heckman model closely resemble those in the conventional probit model, and the small and insignificant coefficient for ϱ indicates that selection bias does not substantially affect the parameter estimates. As expected, the population served strongly predicts selection into the AWWA dataset; reliance on surface sources and location in the West or Midwest, rather than in the Northeast, are also important factors. Utilities in the AWWA dataset are not a random sample of public water utilities nationwide, but unmeasured aspects of the selection process appear not to influence the dependent variable in the rate-choice model.

Robustness

The result for special district governance is robust across a variety of model specifications. Altering the sample to include only utilities with a single residential rate structure did not change the results. Treating the

Table A1.2
Heckman Estimates: Effect of Special District Governance on Adoption of Increasing Block Rates

Variable	Probit model with selection: Outcome equation	Probit model with selection: Selection equation	Probit model
Special district	1.28***		1.39***
	(.46)		(.42)
Temperature	.22***		.27***
	(.07)		(.06)
Special district * Temperature	−.26**		−.27***
	(.11)		(.10)
Precipitation	−.00		−.01
	(.07)		(.06)
Operating ratio	.07		.06
	(.12)		(.11)
Proportion retail sales	1.25***		1.34***
	(.43)		(.44)
Urban	.20		.13
	(.16)		(.16)
Population		.48***	
		(.04)	
Surface water	−.00	.31***	.01
	(.19)	(.07)	(.16)
Purchased water	.31	−.08	.29
	(.21)	(.09)	(.20)
West	.64**	.56***	.46*
	(.30)	(.09)	(.25)
Midwest	−.12	.48***	−.18
	(.29)	(.10)	(.23)
South	−.19	.10	−.39
	(.32)	(.09)	(.27)
Constant	−2.79***	−6.71***	−2.93***
	(.87)	(.39)	(.67)
ϱ		−.05	
		(.15)	
N selected		401	
N total		22,118	427
Pseudo R^2			.17

Notes: Robust standard errors in parentheses. *$p < .10$, **$p < .05$, ***$p < .01$ (two-tailed).

dependent variable as categorical rather than dichotomous and estimating the model with multinomial logit resulted in similar effect estimates for increasing block rates, but null results for the choice between uniform and declining block rates. Several different measures of financial health also produced similar effect estimates and did not influence the finding for governance structure.

It is possible that the differences we see in the adoption of an increasing block rate do not reflect important differences in the underlying pricing system. A utility might adopt a rate structure that is nominally progressive but establish few rate blocks or little variation in price between them, thus approximating a uniform rate structure. Characterizing rate structures by the number of tiers and by the price ratio between the highest and lowest tiers, I found no evidence that governing structure affects the specific characteristics of a progressive pricing system.

To test for potential endogeneity in the choice of governing structure, I estimated the main model by including a variable measuring the number of years since government incorporation. If policy choice can be explained by factors related to institutional design, then more recently formed governments should be more likely to adopt increasing block rates. The reported results hold when age of government is included in the model, and age has no effect on the water-policy choices made by either specialized or general-purpose governments. I also found no evidence that state political factors affected rate-structure choice. To estimate the impact of state political context, I included in the model the state ideology scores that Robert Erikson, Gerald Wright, and John McIver (1993) developed by compiling CBS/*New York Times* exit polls conducted between 1976 and 1988. State ideology had no effect on policy adoption. Furthermore, although I have no theoretical reason to expect dependence among utilities located within the same state, as a check I ran the model with standard errors clustered by state. Clustering standard errors widened confidence bands only slightly, and the institutional difference between special districts and general-purpose governments remained statistically significant at the median temperature.

In addition to these checks on the model's specification and functional form, I employed another utility survey to assess the robustness of the findings more generally. As noted earlier in this appendix, the AWWA sample is not designed to represent the universe of public water utilities. The Heckman model helps account for the overrepresentation of large utilities and those that rely on surface sources, but there may remain unspecified sources of bias in sample selection. To confirm that the results would hold with a more representative sample of water utilities, I tested a reduced form of the model using data from the 1995 Community Water System (CWS) survey conducted by the EPA (U.S. EPA 1997a, 1997b). The EPA periodically gathers data from a national stratified random sample of water utilities. Response rate for the 1995 survey was 54 percent, and the agency generated sampling weights to allow estimation of population characteristics.

Unfortunately for the current purpose, responding utilities were guaranteed anonymity, making it impossible to introduce variables from outside the dataset and carry out the full analysis. Instead, I used EPA region as a proxy for climate data. I determined from the AWWA data that utilities in EPA regions 4, 6, and 9—an area that encompasses the Sunbelt states—scored substantially higher on the temperature variable than did utilities in other regions. The reduced model includes this regional proxy, governance type, and the interaction between them, as well as control variables measuring water supply source, operating ratio, and proportion retail sales. Marginal effects generated from the two models differ in magnitude, but the direction and substantive interpretation are the same. In the original AWWA data, the marginal effect of water district governance in non-Sunbelt states is 0.33, compared to 0.19 in the EPA data. Both of these effects are significant with 95 percent confidence. Neither of the datasets produces a significant effect for governance type in Sunbelt states where water is a more severe public problem. Although the regional proxy is a blunt instrument for measuring problem severity, these findings increase confidence that the findings reported in the chapter text persist with a more representative sample of public water utilities.

Appendix 2
Explanation of Data and Model, Chapter 4

Data

Data on connection fees come from biannual, odd-year utility surveys conducted between 1991 and 2003 by the consulting firm Black & Veatch. The surveys document one-time connection fees charged to new customers to cover the cost of facilities necessary to serve new development; the fees do not include on-site costs that may apply for service lines, meters, and meter installation. Black & Veatch collects data by utility and municipality; if a water district's connection fee varies across municipalities in its service area, the value of the dependent variable is the average of fees charged by the utility. Fees for all years are in constant 2000 dollars. The sample consists of an unbalanced panel of 316 utilities observed over seven time periods, but findings are robust when restricting the analysis to a balanced panel of 184 utilities.

In most cases, utility names combined with knowledge of general enabling legislation for water districts in California were sufficient to code utilities as overseen by special districts or by general-purpose governments. I consulted the *Census of Governments,* local government financial data from the state controller's office, and utility Web sites to rule on questionable cases. Control variables for utility characteristics come from several sources. Information on a utility's customer base and its water prices come from the Black & Veatch surveys. The measure for monthly residential water charge is a sum of the service charge and the commodity charge calculated using the inside city rate. Data were collected based on a standard meter size and usage level and thus do not reflect true water consumption. The charges for all years are in constant 2000 dollars. Data on indebtedness come from annual reports on local government finances produced by the California State Controller's Office for fiscal year 1998–1999.[1] The EPA's Safe Drinking Water Information System provided information on water supply sources.

Assembling data on the characteristics of utility constituencies required aggregating census-block-level and block-group-level demographic and political data up to the boundaries of the cities and water districts in the sample.[2] Variables measuring population growth, urban population, and multiunit housing are calculated from U.S. Census count data from SF1 files or sampled count data from SF3 files. The income variable measures the median across block-group medians in the jurisdiction. Demographic data for 1991 and 1993 observations come from the 1990 census and use 1990 census block boundary shapefiles; measurements for observations from 1995 forward are based on blocks and data from the year 2000. The growth variable measures the rate of population increase in the jurisdiction between 1990 and 2000. Population estimates are unavailable at the special district or census-block level for years between the decennial census counts, preventing use of a time-variant measure of population growth.

The political ideology measure relies on block-level vote returns compiled by the California Statewide Database (SWDB). As part of its duties as California's redistricting database, the SWDB takes precinct-level electoral returns and disaggregates them to the census block for each election. To construct the ideology indicator for this analysis, I used block-level data on votes cast in the top-ticket race of the previous election and calculated the proportion of the jurisdiction's vote cast for the Democrat. For example, a 1993 observation measures ideology by the proportion of the jurisdiction's voters casting a ballot for Bill Clinton the previous November, and a 1999 observation takes the proportion voting for Gray Davis in the 1998 California gubernatorial election.

Cross-sectional climate measures were constructed by overlaying climate maps produced by the NCDC onto maps of utility location, as described in appendix 1. Time-variant measures of weather conditions used divisional climate data from the NCDC and the NOAA-CIRES Climate Diagnostics Center. Annual climate data are available only at the level of California's seven regional climate divisions, not in the more detailed CLIMAPs that show long-term averages. For each of California's seven regional climate divisions, I collected data on means and standard deviations of annual temperature and precipitation for the period from 1971 to 2000. I then calculated annual departure from those means, measured in numbers of standard deviations. Positive values are hotter and wetter than normal, and negative values are cooler and dryer. The results presented here use a two-year lag in the divisional departures from normal temperature and precipitation in order to provide time for a policy response to changes in water salience.

The final measure of climate conditions used for this chapter is the Palmer Hydrological Drought Index (PHDI), a meteorological drought index used to assess the long-term severity of dry or wet spells of weather. I included this measure to capture large-scale variation in issue salience that might not be evident in the more detailed indicators. The PHDI quantifies hydrological impacts of drought such as reservoir and groundwater levels that take longer to develop and rectify. Possible values on the index range from -7 to 7; 0 represents normal conditions, and negative values indicate drought. Measured at the statewide level, this indicator varies only across time, not by utility, and the model includes values for a two-year lag.

Model

Given that the main explanatory variables of interest, utility governance and population growth, are time invariant within the sample, the model primarily explains between-unit differences in impact fees rather than within-unit change over time. Panel data still provide some advantage over a simple cross-sectional analysis: a panel analysis assures that observed relationships persist longer than the single moment in which we collect data for a cross-sectional analysis, and it allows covariates to vary over time, capturing dynamic change in other factors that contribute to fee levels. Pooling panel cross-sections can also produce more efficient effect estimates than looking at a single cross-section (Stimson 1985). Nevertheless, pooling panel data introduces a number of methodological problems, including heteroskedasticity and serially correlated errors. Customary solutions for these problems are not appropriate for this analysis: first differences and fixed effects are not an option when time-invariant explanatory variables are an important element in the model, and including a lagged dependent variable focuses attention on short-term dynamics and may produce biased estimates that understate the influence of substantive independent variables (Achen 2000).

The effect estimates featured in chapter 4 come from a Prais-Winsten generalized least-squares regression that corrects for first-order autocorrelation.[3] The model includes year dummies to absorb any remaining

Table A2.1
Panel Data Estimates: Effect of Special District Governance on Water Impact Fee Levels

Variable	Prais-Winsten estimates		Between-effects estimates	
Special district	.621	(.266)**	.675	(.247)***
Growth	.865	(.398)**	.852	(.604)
Special district * Growth	−1.583	(.623)**	−1.594	(.778)**
Climate variables:				
Temperature	.085	(.090)	.162	(.101)
Precipitation	.191	(.083)**	.201	(.086)**
Temperature departure	.090	(.052)*		
Precipitation departure	−.056	(.060)		
Drought index	.009	(.028)		
Demand variables:				
Income	.019	(.006)***	.016	(.007)**
Multiunit housing	−2.490	(.800)***	−2.744	(.854)***
Urban	−.054	(.171)	−.046	(.289)
Democrat	−.370	(.441)	1.235	(.753)
Utility variables:				
Water charge	.019	(.007)***	.033	(.009)***
Water debt per capita	.218	(.119)*	.258	(.110)**
Population served	.150	(.052)***	.176	(.090)*
Surface water	.521	(.178)***	.326	(.197)*
Constant	−1.744	(.832)**	−3.228	(1.235)***
Number of observations	1,925		1,925	
Number of groups	316		316	
R^2 (overall)	.133		.236	
R^2 (between)			.285	
ϱ	.839			
Durbin-Watson (original)	.293			
Durbin-Watson (transformed)	1.941			

Notes: Standard errors in parentheses, clustered by utility for the Prais-Winsten estimates. Year fixed effects in the Prais-Winsten model not shown. *$p < .10$, **$p < .05$, ***$p < .01$ (two-tailed).

time dependence. Standard errors are adjusted for the clustering of observations by utility. Coefficients from the complete model appear in table A2.1. As a check on the model's robustness, the table also presents estimates from a between-effects regression on the average values of all variables for each utility over time. By focusing exclusively on differences between units, the between-effects method loses some of the information contained in a panel dataset, but it maintains the focus on the time-invariant measures that are of primary interest. The between-effects model does not include year dummies or the lagged climate variables. As table A2.1 demonstrates, the between-effects estimates support the results from the Prais-Winsten model.

Including years since government incorporation in the model helps address the possibility that political actors make a simultaneous choice about institutional design and water-revenue structures. If special districts get established in order to shift the cost of growth to incoming residents, we should expect more recently formed districts to be particularly reliant on impact fees. In fact, government age has a positive effect on impact fees that is weakly significant ($p < .08$). It is older governments that charge higher fees. A split sample analysis suggests that this effect is exclusive to municipal governments. Among cities, a one-year difference in age produces an estimated $8 increase in fees ($p < .06$). Among special districts, age of government does not have a significant impact on fee levels, and controlling for years since government incorporation in the full model produces a modest increase in specialized governance's effect. Missing data on incorporation dates reduce the sample size for this analysis, so age does not appear in the model reported in the main text.

Appendix 3
Explanation of Data and Model, Chapter 5

Data

The source for data on water districts' boundary flexibility is state law. Starting with a sample of twenty-one states that are most reliant on special districts for water provision, I compiled a list of all water district types in each state using information from the *Individual State Descriptions* volume of the 2002 *Census of Governments* (U.S. Census Bureau 2005b). I then consulted state statutes to code the rules for changing district boundaries. Some states have common boundary-change rules that apply to all special districts. In other cases, boundary rules appear in the statutes enabling each district type. Using information contained in district names and in state and individual district Web sites, I coded each district by type and attached the relevant set of rules. The analysis omits districts formed under specific legislation and districts with boundaries that must be contiguous with a city or county. These omissions removed four states from the sample, producing a dataset containing 1,383 water districts in seventeen states. The seventeen states included in this analysis contain two-thirds of the water supply special districts in the United States.

Data on the dependent variables and on most of the control variables come from the 2002 *Census of Governments* (U.S. Census Bureau 2002a; U.S. Census Bureau 2005a). The proportion of special district governing-board members that are elected comes from the 1992 *Census of Governments* (U.S. Census Bureau 1994). Data on city annexation rules was compiled by the ACIR (1992). Climate maps produced by the NCDC are the source for temperature and precipitation conditions, with data compiled using the process described in appendix 1.

Model

The analysis uses complementary log-log models to estimate the effects of boundary flexibility on the likelihood that a water district engages in interlocal cooperation. The functional form accounts for the rare occurrence of intergovernmental contracts: among the sampled water districts, 6 percent participated in revenue contracts, and 4 percent participated in expenditure contracts.[1] The asymmetric complementary log-log link is appropriate for this type of binary distribution in which positive responses are rare. The models cluster observations according to the state laws that set rules for boundary change. Where a single law governs boundary change for all district types in a state, the state's observations fall into a single cluster. The dataset includes twenty-three clusters of water districts in seventeen states. A likelihood ratio test supports treating the revenue and expenditure models as independent equations.

The result showing a positive relationship between boundary restrictiveness and revenue agreements is robust to estimation using other link functions, including probit and the linear probability model. The negative result for interlocal expenditures does not pass significance tests at the $p < .1$ level when estimated using different models, so the result receives less attention in the text. Table A3.1 reports coefficients from the models that generated probabilities displayed in the text.

In addition to the models shown here, I also estimated an additional expenditures model including two variables intended to measure the demand for a water contract based on a special district's system needs. First, I included years since government incorporation both to address potential endogeneity of institutional choice and to account for older special districts' having infrastructure that is more likely to be failing or inadequate to meet constituent needs. In addition, I included a dichotomous variable indicating whether the district had any reported violations of the Safe Drinking Water Act between 1997 and 2001. Data on violations of the drinking water standard come from the EPA's Safe Drinking Water System/Federal Version, a system with known data reliability problems based in large part on variability in the quality of data reported by states. Missing data on the two variables reduces the sample size for the analysis by 40 percent. Both variables have a small and insignificant effect on expenditure contracts, but the small sample size eliminates the significance of the boundary rules coefficient, whether the additional variables appear in the model or not.

Table A3.1
Complementary Log-Log Estimates: Effect of Boundary Flexibility on Interlocal Cooperation

	Revenue agreements		Expenditure agreements	
Boundary-change rule index	.68	(.19)***	−.65	(.38)*
Contiguity requirement	−.96	(.35)***	−.60	(.56)
Fiscal variables:				
Current expenditures (log)	.12	(.15)	−.11	(.11)
Debt finance	.33	(.16)**	.27	(.27)
Long-term debt (log)	.14	(.10)	.08	(.04)**
Property taxes per capita	.18	(.37)	.35	(.16)**
Intergovernmental variables:				
County local governments	−.00	(.00)	.00	(.00)
Multicounty special district	.19	(.16)	.52	(.32)*
Common boundaries	.68	(.20)***	−.59	(.60)
Proportion spending on water	1.06	(.70)	1.76	(.71)**
City annexation rules	−.18	(.14)	.29	(.21)
District formations	−.26	(.10)***	.03	(.07)
Institutional variable:				
Proportion elected	−.54	(.34)	1.32	(.56)**
Problem severity variables:				
Precipitation	.04	(.13)	−.00	(.09)
Temperature	−.08	(.06)	−.21	(.09)**
Population growth	−1.14	(.97)	.48	(.27)*
Constant	−6.13	(.84)***	−3.86	(1.17)***
N	1,393		1,393	
Log pseudolikelihood	−286.59		−221.16	

Notes: Standard errors in parentheses, clustered by special district type within a state. ***$p < .01$, **$p < .05$, *$p < .10$ (two-tailed).

Notes

Chapter 1

1. Omitted from this analysis are the many thousands of dependent districts, or special districts that are subordinate to another governmental entity. Dependent districts are overseen by the government that established them or by a board appointed by that government, and their financial transactions often require review and approval. Chapter 2 includes further discussion defining special districts and distinguishing them from other units of government.

2. The legal doctrine that defines local governments as creatures of the state is known as Dillon's Rule, articulated by Iowa Supreme Court justice John Dillon in 1868.

3. Bollens was using a phrase that earlier had been applied to counties.

4. See Foster 1997, 15–20, for a concise history of special district emergence in the United States.

5. Nancy Burns (1994) and Kathryn Foster (1997) report mixed evidence on the effect of annexation laws and tax and debt limits. For studies of the impact of state laws regulating local government on special district formation, see Austin 1998; S. Bollens 1986; Carr 2006; MacManus 1981; and McCabe 2000. See also ACIR 1964 and Walsh 1978 on the use of independent and dependent special districts to circumvent local debt limits.

6. Revenue bonds were first used by the city of Spokane, Washington, in 1897. The city sold bonds to finance the extension of its water-supply system and pledged revenues to be received from water users as repayment on the bonds and interest (ACIR 1964). Revenue bonds are now a standard tool that special districts and other local governments use to pay for infrastructure development and extension.

7. Results from quantitative analysis of district formations nationwide are mixed. See Burns 1994 and Foster 1997.

8. See Foster 1997 and C. Berry 2007 for evidence on the spending effects of specialized governance. For a test of the hypothesis that special district governance produces higher levels of corruption, see Meier and Holbrook 1992.

9. For studies of distributional equity in urban-services provision, see Cingranelli 1981; B. Jones 1981; Koehler and Wrightson 1987; Levy, Meltsner, and Wildavsky 1974; Lineberry 1977; and Mladenka 1980, 1989.

10. Sarah Elkind (2000) offers a compelling account of the formation of regional water districts in Boston and Oakland.

11. Calculated from responses to the 1999 Financial and Revenue Survey conducted by the AWWA (1999).

12. For more on the colorful and sometimes tragic history of these projects, see Fradkin 1996; Hundley 2001; Kahrl 1982; Pisani 2002; Reisner 1986; and Worster 1992.

13. The U.S. Geological Survey defines *public water supply* as water withdrawn by public and private suppliers that furnish water to at least twenty-five people or have a minimum of fifteen connections. Estimates of the percentage of Americans obtaining drinking water from a public supply in 2000 range from 85 percent (Hutson, Barber, Kenny, et al. 2004) to 90 percent (U.S. EPA 2002c). The rest self-supply, typically with private wells.

14. In-stream use for hydroelectric power and environmental protection is not included in these estimates.

15. The State of Washington attempted to establish priority for drinking water uses and greater certainty over freshwater allocations when it passed the Municipal Water Law in 2003. At the time of this writing, the controversial law is still under court review.

16. Using data from a 2000 survey of community water systems, the U.S. EPA estimates that residential use accounts for 66 percent of retail water deliveries. This estimate excludes wholesale deliveries and does not account for system losses. Residential connections account for 91 percent of total retail connections, each residential user consuming less than a commercial or industrial customer (U.S. EPA 2002c).

17. The 15 percent of public water-system withdrawals shown in figure 1.2 that goes to "public use and losses" includes water lost in treatment and distribution systems as well as municipal uses such as firefighting, street washing, and service to parks and public swimming pools.

18. The per capita public water withdrawal data in figure 1.3 are based on the overall U.S. population, not on the population receiving water from a public system.

19. Data sources: Gallup Organization, Gallup Poll (April 3–9, 2000; March 5–7, 2001; March 4–7, 2002; March 3–5, 2003; March 8–11, 2004; March 13–16, 2006; March 11–14, 2007; March 6–9, 2008). Data retrieved from the iPOLL databank, the Roper Center for Public Opinion Research, University of Connecticut.

20. Data sources: Gallup Organization, World Values Survey (September 1995); Gallup Organization, Gallup/CNN/USA Today Poll (April 13–14, 1999); Gallup Organization, Gallup Poll (April 3–9, 2000). Data retrieved from the iPOLL databank, the Roper Center for Public Opinion Research, University of Connecticut.

21. For more on the global trend toward decentralized water management, see Wescoat and White 2003.

22. International organizations are now turning their attention to developing local capacity to implement these solutions. A key component of the 2009 World Water Forum is a focus on strengthening local authorities so they can carry out their responsibilities under decentralization.

23. The National Research Council Water Science and Technology Board (2001, 2004) has identified social science research on political and economic institutions for water management as a critical element on the agenda for water research in the twenty-first century.

24. For analysis of water privatization in the United States, see Gleick, Wolff, Chalecki, et al. 2002; Haarmeyer and Coy 2002; National Research Council 2002b; Spulber and Sabbaghi 1998; U.S. Conference of Mayors Urban Water Council 1997; Werkman and Westerling 2000; and Wolff and Hallstein 2005. Stronger advocacy statements appear in Barlow and Clarke 2002; Luoma 2002; and Snitow and Kaufman 2007.

25. Organizational diversity exists among privately owned water systems just as it does among publicly owned systems. Just 27 percent of privately owned water systems operate as for-profit businesses. One-third are nonprofit entities, and the remainder are ancillary water systems in which water supply is ancillary to the primary business. Mobile home parks are the most common example of an ancillary system (U.S. EPA 2002c). The process of privatization can also take a variety of forms, including outsourcing services; contracting the operation of existing plants; designing, building, and operating contracts for new facilities; and selling public assets to investor-owned companies. The outright sale of assets is uncommon (National Research Council 2002b).

26. The category "water districts" includes single-function special districts that manage water supply and multifunction districts that combine water supply with sewerage or natural-resource functions such as flood control.

27. For discussion of the growing interrelationship between water and land-use planning, see Page and Susskind 2007; Shigley and Krist 2002.

28. In some areas, consumers may have the option of drilling a well, but only commercial consumers are likely to pursue this high-cost endeavor. The overdrafting of many groundwater aquifers adds to the risk in abstaining from public water service. Bottled water is an impractical and expensive alternative for most uses.

29. Figure 1.5 indicates that the water districts in North Dakota currently are inactive.

Chapter 2

1. On representation, see Banfield and Wilson 1963; Schumaker and Getter 1977; Welch 1990; and Welch and Bledsoe 1988. On policy effects, see Clingermayer and Feiock 2001; Karnig 1975; and Lineberry and Fowler 1967. In contrast, Morgan and Pelissero 1980; Sharp 1986; and Wolfinger and Field 1966 found reform structures to be less consequential.

2. The U.S. Census Bureau's definition includes several other criteria for special district independence, specifying elements of fiscal and administrative autonomy in greater detail.

3. I also required that special districts have autonomy in making policy decisions, particularly in the areas of my empirical tests. Coding decisions for the empirical analyses were based on self-reported governing structures in water-utility surveys supplemented with data from local government Web sites, the U.S. Census Bureau, and review of state laws defining special district scope and responsibilities. The precise criteria for determining governance type ultimately made little difference for my findings. Rarely was there any question about how to treat an individual government, and the results presented in this volume would still hold if multiple definitions and coding rules were used.

4. These figures are calculated from 2002 *Census of Governments* (U.S. Census Bureau 2002a) data on the subset of special districts that report information about the territory they serve. Response rates for this survey item are low, and data are available for fewer than two-thirds of special districts.

5. A special district's boundaries are sometimes set for overtly political reasons. Many districts in Texas have been designed specifically to exclude any existing residents so that developers' agents can move onto the property and cast the single vote necessary to pass a bond-financing infrastructure development, passing the burden of repayment on to future homeowners (Egerton and Dunklin 2001; Perrenod 1984).

6. See, for example, Bawn 1995; Epstein and O'Halloran 1999; Lowi 1969; McCubbins, Noll, and Weingast 1987, 1989; McCubbins and Schwartz 1984; and Weingast and Moran 1983.

7. Douglas Porter, Ben Lin, and Richard Peiser (1987) argue that an advantage of special districts is their political insulation, allowing districts to carry out their activities free from political considerations. Carolyn Bourdeaux (2005) treats special districts' political insulation as hindering effective governance because officials will spend more time pursuing projects that are not politically viable. John Bollens (1957) and Robert Smith (1964), like the ACIR, contend that keeping a function "out of politics" is impossible and perhaps undesirable.

8. For representative works, see ACIR 1964; Committee for Economic Development 1966, 1970; V. Jones 1942; and Wood 1958, 1961.

9. For the effect of election timing on turnout in municipal elections, see Hajnal and Lewis 2003.

10. See *Salyer Land Company v. Tulare Lake Basin Water Storage District*, 410 U.S. 719 (1973), and *Ball v. James*, 451 U.S. 355 (1981).

11. *Responsiveness* is defined here as the degree to which governments deliver policies that are consistent with the preferences of a majority of their constituents. A responsive government recognizes and deciphers constituent demands and sets policy to satisfy those demands. It is guided by the median resident's preferences and does not display a bias toward minority interests with more intense preferences and more resources to invest in lobbying. Motivated by the median-voter model and incumbent politicians' reelection goals (Black 1958; Downs 1957), this definition is consistent with the public opinion and public policy literatures on policy responsiveness (Bartels 1991; Gerber 1996; Gilens 2005; B. Jones 1973; Miller and Stokes 1963; Page and Shapiro 1983; Shumaker and Getter 1977).

12. See ACIR 1964 for a particularly strong statement about the effects of specialized governance on interlocal coordination. Robert Dahl and Edward Tufte argue that "indefinite specialization, like indefinite proliferation of units, would lead sooner or later—and may already have—to so much fragmentation as to create insuperable problems of coordination" (1973, 142).

13. See Egerton and Dunklin 2001; Florida Office of Program Policy Analysis and Government Accountability 1995; Little Hoover Commission 2000; and New York State Comptroller's Office 2007.

14. Representative works include Bish 1971; Hawkins 1976; McGinnis 1999; Ostrom, Bish, and Ostrom 1988; and Ostrom, Tiebout, and Warren 1961.

15. Richard Wagner and Warren Weber (1975) modeled local governance as monopolistic rather than competitive, but they reached the same conclusion that specialized governance typically operates more efficiently than a single general-purpose government offering multiple services.

16. Studies of the efficiency effects of fragmentation include Forbes and Zampelli 1989; Oates 1989; Schneider 1989; and Zax 1989. For a survey of the empirical literature, see Dowding, John, and Biggs 1994.

17. Steven Erie (2006) has challenged this characterization, arguing that the institutional design of southern California's Metropolitan Water District makes it more effective in satisfying public demands for economic development and water reliability.

18. The broader framework for institutional analysis developed by leading public choice scholars emphasizes the contingency of institutional effects (E. Ostrom 2005). However, propositions about the comparative responsiveness of specialized and general-purpose governments are not expressed as conditional hypotheses (Ostrom, Bish, and Ostrom 1988).

19. The assumption of a single-dimension policy space implies that voters are evaluating candidates and policies according to a single, common criterion.

20. The Supreme Court's decisions in *Salyer* and *Ball* (see note 10 in this chapter) focused on the narrow, "special limited purpose" of the agricultural water

districts in question and the disproportionate impact of district decisions on landowners. Because the districts did not carry out "normal functions of government," the Court ruled that they could depart from the one-person, one-vote standard for local government set in its decision for *Avery v. Midland County*, 390 U.S. 474 (1968). The Court has not defined what constitutes a "special limited purpose" or "normal functions of government," but, in practice, land-based voting rules apply in a relatively small number of irrigation and natural resources districts. For analysis of the Court's approach to voting rights as applied to local governments, see Briffault 1993.

21. Tanya Heikkila and Kimberly Isett (2007) offer evidence to suggest that among citizens active in local government, variation exists across communities in levels of knowledge regarding special district organization and activities.

22. Typical public opinion measures of issue salience may in fact capture objective conditions in an issue area rather than the issue's importance. Christopher Wlezien (2005) shows that much of the variation in responses to the "most important problem" question that researchers commonly use to measure salience can be attributed to problem status, telling us little about the issue's importance to survey respondents.

23. One might argue that all policy questions that involve public spending inherently cross functional boundaries because of competition among functions and governments for residents' tax dollars. Findings in the public opinion literature suggest that voters rarely recognize or account for these trade-offs when they develop preferences about any one spending decision.

24. Forty percent of the rating that the insurance industry assigns to evaluate the efficacy of a community's fire protection is based on an appraisal of water-supply capacity (Hickey 2002).

25. On committee systems in legislatures at different levels of government, see, for example, Cooper 1971; Pelissero and Krebs 1997; and Rosenthal 1990. All decision-making bodies, both specialized and multifunction, may have difficulty addressing policy proposals that involve multiple evaluative dimensions. Majority support for multidimensional proposals is often unstable. One of the core functions of committee systems is to convert complex, multidimensional issues into single-dimensional policy proposals for voting by the legislative body (Shepsle 1979; Shepsle and Weingast 1987). Because a special district's issue scope is limited by design, many of the policy questions a district will address fit the one-dimensional framework.

26. Calculated using data from the employment phase of the 2002 *Census of Governments* (U.S. Census Bureau 2004).

27. See Feiock 2004 and Feiock and Scholz 2009 for further discussion of institutional collective action. Richardson Dilworth (2005) also depicts metropolitan fragmentation as a form of collective action problem.

28. For example, Jameson Doig (1983) highlights the concern that political insulation extends to insulation from other elected officials, potentially interfering with the development of cooperative relationships.

29. For a recent review of the debate and the empirical evidence, see Altshuler, Morrill, Wolman, et al. 1999.

30. The last time the *Census of Governments* reported data on incorporation dates for a large number of special districts was 1987; where possible, I used information on water district and state Web sites to fill in more recent formation dates. As of 1987, the median incorporation date for the 3,081 independent water districts in the census dataset was 1965 (U.S. Census Bureau 1990).

31. For more information on these climate variables, see appendix 1.

Chapter 3

1. This explanation of water pricing by nineteenth-century municipal utilities relies in large part on Crocker and Masten 2002.

2. Where urban water utilities provide service to agricultural customers, the pricing structure may differ markedly from the pricing structures for other customer classes.

3. This chapter addresses only the fees on water consumption. Most water bills include a fixed service charge to cover customer costs such as metering and billing as well as a commodity charge on consumption. In addition, utilities may impose impact fees and other charges for extending and expanding service. Impact fees are the focus of analysis in chapter 4.

4. From 1970 through 1984, water prices rose at a rate slightly lower than inflation. After 1984, water-price increases consistently exceeded rates of inflation. Sewer and cable television prices rose at approximately the same rate as water, whereas other utilities (gas, electricity, and telephone) experienced lower rates of increase (Beecher and Mann 1997; U.S. EPA 1997a).

5. Water is very affordable in the United States when compared to its cost in the rest of the world, but the Organization for Economic Cooperation and Development (OECD) notes that reliance on median household income as the leading affordability indicator might cause U.S. water affordability to be overestimated. In predicting the compliance costs of proposed drinking water regulations, the EPA has determined that water rates should not exceed an affordability threshold of 2 percent of median income. Almost all water systems have rates that fall within this threshold, but the OECD estimates that 10 to 25 percent of households in the United States pay more than 2 percent of their income on water (OECD 2003, 45).

6. Data are not available showing residential water-rate structures over time for a representative sample of utilities nationwide. The data in figure 3.1 were compiled by the OECD (1999) from biennial rate surveys conducted by the Ernst & Young and Raftelis consulting groups. These surveys concentrate on utilities that serve large metropolitan areas, which are more likely to employ increasing block rates.

7. The true marginal cost of water cannot be calculated due to the existence of multiple sources of supply, the escalating and uncertain cost of acquiring future

supply, the large upfront capital costs of system expansion, and the joint consumption that causes marginal cost to vary based on time of use. Moreover, for public water systems there are legal and political constraints on the high profits that marginal pricing would produce during certain periods (Pint 1999). An increasing block rate structure allows utilities to charge a price that approximates marginal cost for unnecessary water uses while meeting politically imposed zero-profit constraints (Olmstead, Hanemann, and Stavins 2005).

8. The effect of block pricing on the elasticity of drinking water demand is a matter of some debate, with varying estimates based on model specification and the type of data examined. A meta-analysis of more than three hundred price-elasticity estimates found that elasticity is higher among consumers facing increasing block rates, meaning that these consumers' demand is more responsive to a change in price (Dalhuisen, Florax, de Groot, et al. 2003). A recent large empirical study (Olmstead, Hanemann, and Stavins 2005) confirmed this finding. The study's authors do not rule out the possibility that differences in elasticities across rate structures may be attributable to endogenous policy choices of the kind examined here. In other words, increasing block pricing may not necessarily reduce water demand, but its benefits for economic efficiency remain unchallenged.

9. For example, see California Public Utilities Commission 2005; Elfner and McDowell 2004; Maryland Department of the Environment 2003; and U.S. EPA 2002a.

10. Following the literature on this type of rate structure, this chapter uses the terms *increasing block*, *progressive*, and *conservation rates* interchangeably.

11. The importance of income in determining residential water consumption is a consistent result throughout the economic literature on urban water demand. Income has both a direct effect and an indirect effect through housing characteristics such as number of bathrooms, home size, and lot size that are strongly correlated with income.

12. Some analysts have expressed concern that increasing block rates are not always progressive in the case of large low-income families, who are more likely than smaller households to consume water in higher-priced tiers (Gomez and Wong 1997; OECD 2003). Some water systems have attempted to address this concern; for example, the Los Angeles Department of Water and Power allows families larger than the average size to petition for a larger allotment in their first tier. The department considered having the rate structure's switch points vary by household size but rejected the plan because it was too difficult to monitor (Mitchell and Hanemann 1994).

13. Outside the United States, increasing block water rates are often called "social tariffs" because of their strong implications for equity (OECD 2003). Relative to other Western democracies' utilities, U.S. and Canadian utilities were late to adopt increasing block pricing.

14. Lifeline rates can take a variety of different forms, including that of an increasing block rate structure (AWWA 1991; Chestnutt, Beecher, Mann, et al. 1997). The 2000 edition of the AWWA rate manual showed greater acceptance

of lifeline rates than had earlier editions. The association previously criticized the rates for departing from cost-of-service principles and clearly stated that rate designers should not institute lifeline rates unless required by legislative mandate (AWWA 1991, 51). The current edition (AWWA 2000) acknowledges the spread of lifeline rates and discusses not only their benefits for customers, but also their potential for lowering a utility's collection costs and nonpayment rates. The manual further notes that lifeline rates can increase public acceptance of an overall rate schedule.

15. Pricing may also be a more equitable approach to conservation than rationing, which forces low-income households to forgo some essential uses (Duke, Ehemann, and Mackenzie 2002). However, cost savings from water conservation that might be induced by block pricing will likely accrue disproportionately to high-income households, especially in the long term.

16. Local officials appear to consider any or all of these public goods when they make the choice to adopt increasing block prices. For example, the City of Santa Barbara switched to a block-pricing schedule during California's severe drought in the late 1980s. The primary goal was to achieve reductions in usage, but city officials also anticipated that wealthier households would continue to consume large amounts of water and thus pay higher water bills (Renwick and Archibald 1998).

17. James Q. Wilson (1980) refers to the dynamics that surround passage of this type of policy as "entrepreneurial politics."

18. The choice of a water-rate structure almost always occurs at the level of the utility governing board—for the cases in this analysis, the water district board or the city or county council—allowing direct measurement of the effects of institutional organization on rate-structure choice.

19. See the influential paper by Allan Meltzer and Scott Richard (1981) for a discussion of the relationship between income inequality and redistributive policy.

20. The authors compiled real data from several water systems into a hypothetical dataset for a single utility.

21. Data come from the 1999 Financial and Revenue Survey conducted by the AWWA (1999). Time-series data on rate structures for a national sample of utilities are not available. Although cross-sectional analysis cannot measure the influence of regional diffusion or highly variable factors contributing to policy adoption (Berry and Berry 1990), it can reveal important stable relationships that explain a policy's existence. Given the temporal stability of the explanatory variables in the model and the relative novelty of the policy under study, time order should not affect the findings' validity.

22. A small percentage of utilities within the sample employ multiple rate structures for residential customers. Most of these utilities combine a flat-fee structure with some other rate structure because not all of their customers have metered connections.

23. The only economic analysis using national data to examine water-rate structure as an endogenous choice found that private utilities were more likely than

public water systems to use declining block structures, but there was no significant difference between ownership types in the probability of adopting increasing block rates (Hewitt 2000). The study did not address the diversity among public governing structures. An analysis of California utilities showed that special districts were more likely than both municipal and private water systems to adopt increasing block rates (Hanak 2005).

24. See Mullin 2008a for evidence that omission of these constituency characteristics does not bias the results presented here.

25. Details on control variables appear in appendix 1.

26. See the report by the National Research Council Water Science and Technology Board (2002a) on models for estimating water use.

27. See appendix 1 for checks on the robustness of these results.

28. I omitted these interactions in the final model to avoid added multicollinearity.

29. Increasing block rates for commercial and industrial users are less common than for residential users, but correlations for increasing block rate use across customer classes are high. Estimating the model for commercial rate systems yields nearly identical results. For industrial rates, specialized governance has a positive impact on progressive pricing, but the effect is not conditional on problem severity, which may be attributable to an economic growth imperative that makes cities and counties less likely to impose high costs on heavy water users. The sample is smaller for analysis of industrial rates, however, so results should be interpreted with caution.

30. Using an original water-utility survey and an alternative measure of water scarcity, Manuel Teodoro (2008) reports evidence confirming the effect of specialized governance and the importance of problem status in conditioning the impact of utility governance.

31. Previous work shows that ward elections tend to produce more low-income city council members (Welch and Bledsoe 1988), suggesting that governments elected by ward might be more likely to adopt progressive rates that promote income redistribution. Evidence that ward elections reduce the likelihood of these rates would suggest a powerful effect for institutional design.

32. In addition to the control variables described earlier, also included in the model for general-purpose governments are measures of the jurisdiction's form of government. These variables isolate the effect of electoral incentives from the effect of reform city management, which may exercise its own independent influence on the adoption of redistributive rates.

Chapter 4

1. There is some confusion regarding the terms *impact fee* and *connection fee*. I use the term *impact fee* to describe only the cost of off-site facilities necessary to produce, treat, and transmit water to new development. This fee is distinct from a connection fee for the on-site costs of establishing a new residential connection

(e.g., service lines from the street to the house, meters, and meter installation). The distinction is often unclear in practice, and many local governments combine fees for on-site and off-site costs (Landis, Larice, Dawson, et al. 1999). The definition I employ makes impact fees for water equivalent to impact fees for local facilities such as schools, roads, and parks.

2. The cases establishing the nexus are *Nollan v. California Coastal Commission*, 483 U.S. 825 (1987), and *Dolan v. City of Tigard*, 512 U.S. 874 (1994).

3. See Altshuler and Gómez-Ibáñez (1993) and Weschler, Mushkatel, and Frank (1987) for further discussion.

4. Impact fees are also more attractive than approving new development without providing for adequate public facilities. This policy choice occurs by default in many communities, and, like funding infrastructure expansion through taxes or user fees, it imposes the cost of new development on new and existing residents alike (Altshuler and Gómez-Ibáñez 1993; Downs 1994).

5. For further discussion and examples, see Baden, Coursey, and Kannegiesser 1999; Downs 1994; Landis, Larice, Dawson, et al. 1999; and Porter 1987.

6. These options receive further attention in chapter 6.

7. One study (Landis, Larice, Dawson, et al. 1999) used data on the full package of impact fees—those imposed by special districts as well as those imposed by the city or county—in a sample of California jurisdictions, but it did not distinguish between fees charged by different governments. The authors had little success identifying the determinants of impact fee adoption, concluding that adoption decisions are ad hoc.

8. Impact fees can be designed to reflect the marginal cost of providing service. For example, lot size is one of the most important determinants of water consumption and cost of water service, so impact fees based on lot size have the potential to promote conservation and economic efficiency. Variable impact fees based on cost of service are rare in practice, however, and the impact fees that do exist may be no more efficient than alternative forms of financing (Baden, Coursey, and Kannegiesser 1999; Speir and Stephenson 2002).

9. The regressive impact of land-use exactions is subject to some dispute (Ihlanfeldt and Shaugnessy 2004).

10. New regionalists are somewhat more optimistic about the potential for regional special districts to coordinate with one another on growth issues. See especially Downs 1994, 173–174.

11. Data on impact fees come from a series of biannual, odd-year utility surveys conducted by the consulting firm Black & Veatch. More details on the data source appear in appendix 2. Fees for all years are in constant 2000 dollars.

12. Many of the variables expected to influence water-pricing decisions are characteristics of a utility's customer base that might affect customers' water demand and their preferences about fees and services. Demographic and political data are unavailable for the special district level, making it impossible to control for constituency characteristics. California is one of the few states that maintain

geographic data on water district boundaries, which allows these variables to be calculated and included in the model.

13. California experienced a 13.8 percent population increase between 1990 and 2000, compared to a 13.2 percent national rate of increase.

14. The courts are still defining what fees are property related under the provisions of Proposition 218. In the key decision *Richmond v. Shasta Community Services District*, 32 Cal. 4th 409 (2004), the California Supreme Court ruled that water impact fees are not subject to Prop 218's vote requirement.

15. The 1982 Mello-Roos Community Facilities Act allows local governments to establish community facilities districts by a two-thirds vote of district residents and to issue tax-exempt bonds for infrastructure with repayment by special taxes. The taxes are borne by all district residents, making them equivalent to property taxes for the purpose of my analysis.

16. Use of impact fees in California occurs under the authority of the Mitigation Fee Act of 1987 (California Government Code §66000–66025), which explicitly states that the legislative body of a local government must enact a new connection fee or a fee increase by ordinance or resolution; the body may not delegate authority for this decision. The act also codifies a nexus requirement that a reasonable relationship exist between a fee's source and its use. Revenues from an impact fee may be used for extending lines and facilities but not to subsidize operating expenses, and they must be kept and administered in an account separate from other revenues.

17. The pattern in impact fee levels is the same if we look just at the balanced sample of utilities reporting fee levels in all time periods.

18. Details on control variables and estimation strategies appear in appendix 2.

19. The 14 percent median growth rate within the sample closely matches California's 13.8 percent population increase between 1990 and 2000.

20. In their study of conservation amendments to general plans in Florida counties, Mark Lubell, Richard Feiock, and Edgar Ramirez (2005) show a similar pattern in which a stronger presence for real-estate interests has opposing effects on amendment adoption for counties with different forms of government.

21. Table 4.1 reports the estimated direct effects of climate variables on impact fee levels. I also tested the possibility of interaction between specialized governance and water scarcity by including in the model interactions with all climate variables, both individually and jointly; none of the interactions had an effect that approached significance.

22. The time trend shown in figure 4.1 suggests that long-term drought may have an effect that is not captured by the annual data.

Chapter 5

1. Empirical evidence on special district boundary changes is scant. The literature on city boundary changes has identified cost savings in service provision as

one motivation for annexation, along with other rationale such as tax-base expansion and local preferences to change the racial balance of the city (Austin 1999; Feiock and Carr 2001; Liner 1990; Liner and McGregor 1996). Efficiency and service considerations are likely to predominate in the case of special district boundary changes. Special district boundaries do not shape perceptions of local political community as city boundaries do, and special districts lack the land-use authority that would allow them to practice exclusionary politics (Danielson 1972). Moreover, it is uncommon for special districts to manage redistributive functions that create the strongest justification for a large tax base. Some special districts do have the authority to impose property taxes, thus creating an incentive to widen their jurisdiction. In general, however, expansion of special district boundaries is most likely a response to changing service demands or to opportunities to achieve policy efficiencies.

2. The potential for cost savings is a dominant reason for contracting by municipalities (Morgan and Hirlinger 1991; Stein 1990).

3. Seasonal variation in demand for water is another important source of slack resources. In climates susceptible to drought, seasonal variation also is an important source of uncertainty about water demand.

4. For a more detailed treatment of special district boundary change and interlocal cooperation within the institutional collective action framework, see Mullin 2009.

5. In a thorough analysis of municipal service arrangements, Robert Stein (1990) found no consistent relationship between a city's annexation authority and its decision to contract for services. Cities rarely have the opportunity to choose between these tools, however, because city boundaries cannot overlap. Annexation requires some supply of annexable land, which is more readily available for water districts than for municipalities.

6. Previous research has shown that a larger supply of potential partners increases the incidence of interlocal agreements (Post 2002). Boundary change may create potential partners not by creating new governments, but rather by situating governments in a position that promotes cooperation.

7. On BexarMet's motivations, see Allen 2005.

8. It is a matter of dispute whether BexarMet had the legal authority to expand its boundaries in the manner it did. Neighboring water systems brought lawsuits charging that BexarMet was not authorized to annex territory outside the boundaries set in its enabling legislation.

9. Research on the effect of state boundary-change rules on city annexations has produced mixed results (Dye 1964; Feiock and Carr 2001; Galloway and Landis 1986; Liner 1990; Liner and McGregor 1996; MacManus and Thomas 1979). Data are not available for special district boundary changes as they are for city annexations, so we cannot measure the impact of boundary rules on special district expansions.

10. Standard wholesale sales of water to another district are not counted as interlocal agreements in the census data. In the AWWA's 1999 Financial and

Revenue Survey (AWWA 1999), 75 percent of U.S. water districts reported at least one wholesale water connection.

11. In some cases, referendum or majority approval is required only upon request. Because local actors always must consider the possibility that someone will request the referendum, these cases are coded as requiring majority approval.

12. Scale reliability as measured by Cronbach's alpha is 0.64.

13. The seventeen states included in this analysis include two-thirds of the water-supply special districts in the United States.

14. Due to data limitations, missing from the model are special district demographic characteristics that might influence local preferences for cooperative agreements. Without reliable, nationwide data on the location of district boundaries, it is impossible to measure the composition of a district's population.

15. Details on control variables and the estimation strategy appear in appendix 3.

16. The only substantive consequences of omitting outstanding debt from the model for revenue agreements are to more than double the size of the coefficient for district expenditures and to increase confidence in rejecting the null hypothesis to the highest level ($p < .001$).

17. The requirement for a public hearing is sufficiently widespread that its inclusion in a multivariate model perfectly predicts the absence of an interlocal agreement. Hearing requirements therefore have been dropped from the models reported in table 5.3.

Chapter 6

1. Some of San Antonio's strategies for keeping up with growing water demand have attracted local opposition for their potentially negative environmental impacts (Glennon 2002).

2. For example, see Macauley 2007 on Currituck, North Carolina; Alexander 2006 on Toms River, New Jersey; Mosely 2006 on Bell Buckle, Tennessee; Manochio 2005 on Roxbury, New Jersey; Nealon 1991 on Brockton, Massachusetts; McCandlish 2006 on Westminster, Maryland; and U.S. EPA 2006 on Frederick, Maryland.

3. All unattributed quotations come from personal interviews I conducted. Interviewees are identified by affiliation but not by name.

4. Quoted in East Bay Municipal Utility District, "Petitioner's Memorandum of Points and Authorities in Support of Petition for Writ of Mandate and Complaint for Declaratory and Injunctive Relief," *East Bay Municipal Utility District v. Contra Costa County*, Superior Court of California, County of Contra Costa, Case no. C93-00235, 9.

5. Quoted in ibid., 12.

6. Prompted by the Dougherty Valley dispute, EBMUD led a decade-long effort to pass state legislation requiring consideration of water availability in local land-

use decisions. For more on the legislation that eventually passed, see Hanak 2005 and Mullin 2007.

7. Town of Danville, City of San Ramon, City of Walnut Creek, City of Pleasanton, Alamo Improvement Association, Sierra Club, Greenbelt Alliance, Preserve Area Ridgelines Committee, Save Our Hills, and Mount Diablo Audubon Association, "Petitioner's Memorandum of Points and Authorities in Support of Petition for Writ of Mandate and Complaint for Declaratory and Injunctive Relief," *Town of Danville, et al. v. Contra Costa County, et al.*, Superior Court of California, County of Contra Costa, Case no. C93-00231.

8. The exception is Ward 2, the ward closest in proximity to Dougherty and Tassajara valleys. Voters in Ward 2 consistently supported progrowth candidates, and the director who won election in 1990 faced no opposition four years later.

9. City of Livermore and Citizens for Balanced Growth, "Opposition to Respondents' Motion for Preliminary Injunction," *Citizens for Balanced Growth v. Alameda County Flood Control and Water Conservation District*, Superior Court of California, County of Sacramento, Case no. 98 CS02670 (consolidated with *City of Livermore v. Alameda County Flood Control and Water Conservation District*, Case no. 98 CS02671), 9.

10. The City of Pleasanton had been one of the petitioners in the cities and community groups' lawsuit against the Dougherty Valley development.

11. City of Livermore and Citizens for Balanced Growth, "Opposition to Respondents' Motion," 3.

12. Town of Danville, "Opening Memorandum of Points and Authorities in Support of Petition for Writ of Mandate," *Sierra Club v. Contra Costa County* (consolidated with *Town of Danville v. Contra Costa County*), Superior Court of California, County of San Joaquin, Case no. CV020073, 11.

13. Ibid., 11, 12.

14. Ibid., 13.

15. Contra Costa County, "Respondents and Real Parties in Interest's Opposition to Danville Petition," *Sierra Club v. Contra Costa County*.

16. The city's electorate approved the UGB in November 2000, with 71 percent voting in favor.

17. South County Resource Preservation Committee, "Petitioners' Reply Brief in Support of Petition for Writ of Mandamus," *South County Resource Preservation Committee v. City of Rohnert Park*, Superior Court of California, County of Sonoma, Case no. 224976, 3, 8.

18. City of Rohnert Park, "Respondents' Brief in Opposition to Petition for Writ of Administrative Mandamus," *South County Resource Preservation Committee v. City of Rohnert Park*, 4.

19. See Baumgarter and Jones 1993 and Schattschneider 1960 for discussion of the dynamics of conflict expansion.

20. For example, a statement on the Friends of the Eel River Web site, the lead petitioner in the suit, regarding SCWA's expansion plan argued that in approving

the project, "the Water Agency failed to consider its impacts on the Eel River, and the suburban sprawl that would result from the increased water supply" (http://www.eelriver.org/engine.php?bit=strategy, accessed September 2008).

21. Elisabeth Gerber and Justin Phillips (2004) find that local voter requirements for new development have a similar impact on land-use outcomes. Because the endorsement of environmental groups influences voter support for development proposals, voter approval requirements force developers to negotiate with environmental groups and to offer public goods in exchange for development rights. See also Gerber and Phillips 2003.

Appendix 1

1. Seasonal or peak pricing can refer to a wide variety of rate structures, making it difficult to code them correctly. The number of utilities relying solely on peak pricing structures is small enough that including them in the analysis does not change the substantive results, regardless of how they are coded.

2. Personal communication with Kurt Keeley of the AWWA, August 2003.

3. Data from the EPA's 1995 Community Water System survey confirm that the population served by a utility is uncorrelated with rate structure.

Appendix 2

1. The financial reports on special districts organize the section on long-term indebtedness by district activity, making it easy to determine what portion of a multifunction special district's debt is attributable to its water activities. The city and county reports do not separate debt by activity, but they usually indicate the purpose of bond issues and state-financed construction. This variable might underreport water debt for general-purpose governments, however, if portions of unspecified or general-purpose bonds have been dedicated to water activities.

2. Data are available at the city and county level for all the variables in the analysis, but I constructed values for these general-purpose jurisdictions using the same process as for water districts so that any measurement error resulting from the data-aggregation process would be distributed among all observations in the sample.

3. The Wooldridge test clearly rejects the null hypothesis of no serial correlation in the errors. The Prais-Winsten procedure is similar to the Cochrane-Orcutt method but maintains the first observation for each unit (Greene 2003).

Appendix 3

1. Only two utilities in the sample reported both expenditures and revenues from interlocal cooperation.

References

Achen, Christopher H. 2000. Why Lagged Dependent Variables Can Suppress the Explanatory Effect of Other Independent Variables. Paper presented at the Annual Meeting of the Political Methodology Section of the American Political Science Association, July 20–22, Los Angeles.

Achen, Christopher H. 1986. *The Statistical Analysis of Quasi-Experiments.* Berkeley and Los Angeles: University of California Press.

Adams, Carolyn. 2007. Urban Governance and the Control of Infrastructure. *Public Works Management and Policy* 11:164–176.

Adding Water for Growth. 2004. *American City & County,* April 1, 54.

Advisory Commission on Intergovernmental Relations (ACIR). 1992. *Local Boundary Commissions: Status and Roles in Forming, Adjusting, and Dissolving Local Government Boundaries.* Washington, D.C.: ACIR.

Advisory Commission on Intergovernmental Relations (ACIR). 1964. *The Problem of Special Districts in American Government.* Washington, D.C.: ACIR.

Alameda County. 1994. *San Francisco Chronicle,* November 10, B8.

Aldridge, James. 2000. BexarMet Pumping Up Growth on South Side. *San Antonio Business Journal,* December 1, 1.

Alesina, Alberto, Reza Baqir, and Caroline Hoxby. 2004. Political Jurisdictions in Heterogeneous Communities. *Journal of Political Economy* 112:348–396.

Alexander, Andrea. 2006. Water Connections Ban Lifted. *Asbury Park Press,* December 21, 1B.

Allen, Elizabeth. 2005. Commissioner Larson Slams BexarMet's Expansion Plans. *San Antonio Express-News,* August 5, 3B.

Altshuler, Alan A., and José A. Gómez-Ibáñez, with Arnold M. Howitt. 1993. *Regulation for Revenue: The Political Economy of Land Use Exactions.* Washington, D.C.: Brookings Institution.

Altshuler, Alan, William Morrill, Harold Wolman, and Faith Mitchell, eds. 1999. *Governance and Opportunity in Metropolitan America.* Washington, D.C.: National Academies Press.

American Water Works Association (AWWA). 2000. *Principles of Water Rates, Fees, and Charges.* Manual M1, 5th ed. Denver: AWWA.

American Water Works Association (AWWA). 1999. *WATER:\STATS: The Water Utility Database, Financial and General Revenue Sections* [Computer file]. Denver: AWWA.

American Water Works Association (AWWA). 1991. *Water Rates.* Manual M1, 4th ed. Denver: AWWA.

Annin, Peter. 2006. *The Great Lakes Water Wars.* Washington, D.C.: Island Press.

Applause, at Last, for Desalination Plant (editorial). 2007. *Tampa Tribune,* December 22, 16.

Arnold, R. Douglas. 1990. *The Logic of Congressional Action.* New Haven, Conn.: Yale University Press.

Austin, D. Andrew. 1999. Politics vs. Economics: Evidence from Municipal Annexation. *Journal of Urban Economics* 45:501–532.

Austin, D. Andrew. 1998. A Positive Model of Special District Formation. *Regional Science and Urban Economics* 28:103–122.

Axelrod, Donald. 1992. *Shadow Government: The Hidden World of Public Authorities—and How They Control over $1 Trillion of Your Money.* New York: John Wiley.

Baden, Brett M., Don L. Coursey, and Jeannine M. Kannegiesser. 1999. *Effects of Impact Fees on the Suburban Chicago Housing Market.* Working Paper no. 93. Chicago: Heartland Institute.

Banfield, Edward C., and James Q. Wilson. 1963. *City Politics.* Cambridge, Mass.: Harvard University Press.

Barlow, Maude, and Tony Clarke. 2002. *Blue Gold: The Battle Against Corporate Theft of the World's Water.* Toronto: Stoddart.

Barnett, Tim P., and David W. Pierce. 2008. When Will Lake Mead Go Dry? *Water Resources Research* 44: W03201, doi:10.1029/2007WR006704.

Bartels, Larry M. 1991. Constituency Opinion and Congressional Policy Making: The Reagan Defense Build Up. *American Political Science Review* 85:457–474.

Baumgartner, Frank R., and Bryan D. Jones. 1993. *Agendas and Instability in American Politics.* Chicago: University of Chicago Press.

Bawn, Kathleen. 1995. Political Control versus Expertise: Congressional Choices about Administrative Procedures. *American Political Science Review* 89:62–73.

Beecher, Janice A. 1995. Integrated Resource Planning: Fundamentals. *Journal of the American Water Works Association* 87 (June): 34–48.

Beecher, Janice A., and Patrick C. Mann. 1997. Real Water Rates on the Rise. *Public Utilities Fortnightly* 135 (July 15): 42–46.

Been, Vicki. 2005. Impact Fees and Housing Affordability. *Cityscape* 8:139–185.

Benson, Matt. 2005. Council Poised to Scrub Water Tiers. *Fort Collins Coloradoan*, May 12, A1.

Benson, Matt. 2004. Council Rethinks Water Charges. *Fort Collins Coloradoan*, February 2, A1.

Benson, Matt. 2003. Hot + Dry = Expensive. *Fort Collins Coloradoan*, August 24, A1.

Bernstein, Fred A. 2005. One Town Stops Time by Turning Off the Water. *New York Times*, October 9, 10.

Berry, Christopher R. 2007. *Piling On: Overlapping Jurisdictions and the Fiscal Common Pool*. Working Paper no. 07.05. Chicago: Harris School of Public Policy, University of Chicago.

Berry, Frances S., and William D. Berry. 1990. State Lottery Adoptions as Policy Innovations: An Event History Analysis. *American Political Science Review* 84:395–415.

Berry, William D. 1979. Utility Regulation in the States: The Policy Effects of Professionalism and Salience to the Consumer. *American Journal of Political Science* 23:263–277.

Besley, Timothy, and Anne Case. 2003. Political Institutions and Policy Choices: Evidence from the United States. *Journal of Economic Literature* 41:7–73.

Besley, Timothy, and Stephen Coate. 2003. Elected versus Appointed Regulators: Theory and Evidence. *Journal of the European Economic Association* 1:1176–1205.

Besley, Timothy, and Stephen Coate. 2002. *Issue Unbundling via Citizens' Initiatives*. Discussion Paper no. 2857. London: Centre for Economic Policy Research.

Bish, Robert L. 1971. *The Public Economy of Metropolitan Areas*. Chicago: Markham.

Black, Duncan. 1958. *The Theory of Committees and Elections*. London: Cambridge University Press.

Blatter, Joachim, and Helen Ingram, eds. 2001. *Reflections on Water: New Approaches to Transboundary Conflicts and Cooperation*. Cambridge, Mass.: MIT Press.

Blomquist, William. 1992. *Dividing the Waters: Governing Groundwater in Southern California*. San Francisco: ICS Press.

Blomquist, William, Edella Schlager, and Tanya Heikkila. 2004. *Common Waters, Diverging Streams: Linking Institutions and Water Management in Arizona, California, and Colorado*. Washington, D.C.: Resources for the Future.

Bluestein, Greg. 2007. Tenn. Town Has Run Out of Water. Associated Press, November 1.

Bollens, John C. 1957. *Special District Governments in the United States.* Berkeley and Los Angeles: University of California Press.

Bollens, Scott A. 1986. Examining the Link Between State Policy and the Creation of Local Special Districts. *State and Local Government Review* 18:117–124.

Bourdeaux, Carolyn. 2005. A Question of Genesis: An Analysis of the Determinants of Public Authorities. *Journal of Public Administration Research and Theory* 15:441–462.

Brazil, Eric. 1992. A New E. Bay Water Fight. *San Francisco Examiner,* March 22, B1.

Briffault, Richard. 1993. Who Rules at Home? One Person/One Vote and Local Governments. *University of Chicago Law Review* 60:339–424.

Brueckner, Jan K. 1997. Infrastructure Financing and Urban Development: The Economics of Impact Fees. *Journal of Public Economics* 66:383–407.

Burns, Nancy. 1994. *The Formation of American Local Governments: Private Values in Public Institutions.* New York: Oxford University Press.

California Department of Water Resources. 1998. *California Water Plan Update.* Bulletin 160-98. Sacramento: California Department of Water Resources.

California Public Utilities Commission. 2005. *Water Action Plan.* Sacramento: California Public Utilities Commission.

Callahan, Mary. 2000. RP Planners Hammer Out 20-Year Growth Plan. *Santa Rosa Press-Democrat,* July 16, B1.

Carr, Jered B. 2006. Local Government Autonomy and State Reliance on Special District Governments: A Reassessment. *Political Research Quarterly* 59:481–492.

Carr, Jered B. 2004. Whose Game Do We Play? Local Government Boundary Change and Metropolitan Governance. In *Metropolitan Governance: Conflict, Competition, and Cooperation,* ed. Richard C. Feiock, 212–239. Washington, D.C.: Georgetown University Press.

Casey, Laura. 2000. RP Plan to Enlarge City Wins Approval. *Santa Rosa Press-Democrat,* July 26, B1.

Chestnutt, Thomas W., Janice A. Beecher, Patrick C. Mann, Don M. Clark, W. Michael Hanemann, George A. Raftelis, Casey N. McSpadden, David M. Pekelney, John Christianson, and Richard Krop. 1997. *Designing, Evaluating, and Implementing Conservation Rate Structures.* Sacramento: California Urban Water Conservation Council.

Cingranelli, David L. 1981. Race, Politics, and Elites: Testing Alternative Models of Municipal Service Distribution. *American Journal of Political Science* 25:664–692.

Clarke, Wes, and Jennifer Evans. 1999. Development Impact Fees and the Acquisition of Infrastructure. *Journal of Urban Affairs* 21:281–288.

Clingermayer, James C., and Richard C. Feiock. 2001. *Institutional Constraints and Policy Choice: An Exploration of Local Governance.* Albany: State University of New York Press.

Committee for Economic Development. 1970. *Reshaping Government in Metropolitan Areas.* New York: Committee for Economic Development.

Committee for Economic Development. 1966. *Modernizing Local Government to Secure a Balanced Federalism.* New York: Committee for Economic Development.

Complete Results of the Election in 8 Bay Area Counties. 1990. *San Francisco Chronicle,* November 8, A19.

Contra Costa County. 1994. *San Francisco Chronicle,* November 10, B8.

Cooper, Joseph. 1971. *The Origins of the Standing Committees and the Development of the Modern House.* Houston: Rice University Studies.

Crocker, Keith J., and Scott E. Masten. 2002. Prospects for Private Water Provision in Developing Countries: Lessons from 19th Century America. In *Thirsting for Efficiency: The Economics and Politics of Urban Water System Reform,* ed. Mary M. Shirley, 317–347. New York: Elsevier Science.

Dahl, Robert A., and Edward R. Tufte. 1973. *Size and Democracy.* Stanford, Calif.: Stanford University Press.

Dalhuisen, Jasper M., Raymond J. G. M. Florax, Henri L. F. de Groot, and Peter Nijkamp. 2003. Price and Income Elasticities of Residential Water Demand: A Meta-analysis. *Land Economics* 79:292–308.

Danielson, Michael N. 1972. *The Politics of Exclusion.* New York: Columbia University Press.

Davis, Linda. 2002. Water Board Approves Alamo Creek Conservation Proposal. *Contra Costa Times,* October 9, A3.

Deason, Jonathan P., Theodore M. Schad, and George William Sherk. 2001. Water Policy in the United States: A Perspective. *Water Policy* 3:175–192.

DiLorenzo, Thomas J. 1981. The Expenditure Effects of Restricting Competition in Local Public Service Industries: The Case of Special Districts. *Public Choice* 37:569–578.

Dilworth, Richardson. 2005. *The Urban Origins of Suburban Autonomy.* Cambridge, Mass.: Harvard University Press.

Dingfelder, Sadie F. 2004. From Toilet to Tap: Psychologists Lend Their Expertise to Overcoming the Public's Aversion to Reclaimed Water. *APA Monitor* 35 (8):26.

Doig, Jameson W. 1983. "If I See a Murderous Fellow Sharpening a Knife Cleverly...": The Wilsonian Dichotomy and the Public Authority Tradition. *Public Administration Review* 43:292–304.

Dowding, Keith, Peter John, and Stephen Biggs. 1994. Tiebout: A Survey of the Empirical Literature. *Urban Studies* 31:767–797.

Downs, Anthony. 1994. *New Visions for Metropolitan America*. Washington, D.C., and Cambridge, Mass.: Brookings Institution and Lincoln Institute of Land Policy.

Downs, Anthony. 1957. *An Economic Theory of Democracy*. New York: Harper and Row.

Doyle, Martin W., Emily H. Stanley, Jon M. Harbor, and Gordon S. Grant. 2003. Dam Removal in the U.S.: Emerging Needs for Science and Policy. *EOS* 84:29–33.

Dresch, Marla, and Steven M. Sheffrin. 1997. *Who Pays for Development Fees and Exactions?* San Francisco: Public Policy Institute of California.

Dubin, Jeffrey A., and Douglas Rivers. 1989. Selection Bias in Linear Regression, Logit, and Probit Models. *Sociological Methods and Research* 18:360–390.

Duke, Joshua M., Robert W. Ehemann, and John Mackenzie. 2002. The Distributional Effects of Water Quantity Management Strategies: A Spatial Analysis. *Review of Regional Studies* 32:19–35.

Dye, Thomas. 1964. Urban Political Integration: Conditions Associated with Annexation in American Cities. *Midwest Journal of Political Science* 8:430–466.

Eger, Robert J., III. 2006. Casting Light on Shadow Government: A Typological Approach. *Journal of Public Administration Research and Theory* 16:125–137.

Egerton, Brooks, and Reese Dunklin. 2001. Government by Developer. *Dallas Morning News*, series of articles, ed. Pam Maples, June 10–December 30.

Elfner, Mary A., and Robin J. McDowell. 2004. Water Conservation in Georgia: Bringing Efficiency into Mainstream Thinking. *Journal of the American Water Works Association* 96:136–142.

Elkind, Sarah S. 2000. *Bay Cities and Water Politics*. Lawrence: University of Kansas Press.

Engstrom, Richard L., and Michael D. McDonald. 1981. The Election of Blacks to City Councils: Clarifying the Impact of Electoral Arrangements on the Seats/Population Relationship. *American Political Science Review* 75:344–354.

Epstein, David, and Sharyn O'Halloran. 1999. *Delegating Powers: A Transaction Cost Politics Approach to Policy Making under Separate Powers*. New York: Cambridge University Press.

Erie, Steven P. 2006. *Beyond Chinatown: The Metropolitan Water District, Growth, and the Environment in Southern California*. Stanford, Calif.: Stanford University Press.

Erikson, Robert S., Gerald C. Wright, and John P. McIver. 1993. *Statehouse Democracy: Public Opinion and Policy in the American States*. New York: Cambridge University Press.

Eu, March Fong. 1990. *Statement of Vote: General Election November 6, 1990*. Sacramento: California Secretary of State.

Evans-Cowley, Jennifer S., Fred A. Forgey, and Ronald C. Rutherford. 2005. The Effect of Development Impact Fees on Land Values. *Growth and Change* 36:100–112.

Farooq, Sajid. 2005. Water Politics Shape Dougherty Valley. *Oakland Tribune*, August 8, Tri-Valley sec. Retrieved from Access World News database.

Feiock, Richard C., ed. 2004. *Metropolitan Governance: Conflict, Competition, and Cooperation.* Washington, D.C.: Georgetown University Press.

Feiock, Richard C., and Jered B. Carr. 2001. Incentives, Entrepreneurs, and Boundary Change: A Collective Action Framework. *Urban Affairs Review* 36:382–405.

Feiock, Richard C., and John T. Scholz, eds. 2009. *Self-Organizing Federalism: Collaborative Mechanisms to Mitigate Institutional Collective Action.* New York: Cambridge University Press.

Field, Christopher B., Linda D. Mortsch, Michael Brklacich, Donald L. Forbes, Paul Kovacs, Jonathan A. Patz, Steven W. Running, and Michael J. Scott. 2007. North America. In *Climate Change 2007: Impacts, Adaptation, and Vulnerability. Contribution of Working Group II to the Fourth Assessment Report of the Intergovernmental Panel on Climate Change,* ed. Martin L. Parry, Osvaldo F. Canziani, Jean P. Palutikof, Paul J. van der Linden, and Clair E. Hanson, 617–652. Cambridge, U.K.: Cambridge University Press.

Fischel, William A. 2001. *The Homevoter Hypothesis: How Home Values Influence Local Government Taxation, School Finance, and Land-Use Politics.* Cambridge, Mass.: Harvard University Press.

Florida Office of Program Policy Analysis and Government Accountability. 1995. *Review of Independent Special Districts That Provide Infrastructure and Services to the Public.* Report no. 95-22. Tallahassee, Fla.: Office of Program Policy Analysis and Government Accountability.

Forbes, Kevin F., and Ernest M. Zampelli. 1989. Is Leviathan a Mythical Beast? *American Economic Review* 79:568–577.

Foster, Kathryn A. 1997. *The Political Economy of Special Purpose Government.* Washington, D.C.: Georgetown University Press.

Fradkin, Philip L. 1996. *A River No More: The Colorado River and the West.* Berkeley and Los Angeles: University of California Press.

Frey, Bruno S., and Reiner Eichenberger. 1999. *The New Democratic Federalism for Europe: Functional, Overlapping, and Competing Jurisdictions.* Northampton, Mass.: Edward Elgar.

Frieden, Bernard J. 1983. The Exclusionary Effect of Local Growth Controls. *Annals of the American Academy of Political and Social Science* 465 (Housing America): 123–135.

Fulton, William, Rolf Pendall, Mai Nguyen, and Alicia Harrison. 2001. *Who Sprawls Most? How Growth Patterns Differ across the U.S.* Washington, D.C.: Brookings Institution.

Galloway, Thomas D., and John Landis. 1986. How Cities Expand: Does State Law Make a Difference? *Growth and Change* 17:25–45.

Gaumnitz, Lisa, Tim Asplund, and Megan R. Matthews. 2004. A Growing Thirst for Groundwater. *Wisconsin Natural Resources Magazine* (June–July). Retrieved from http://www.wnrmag.com/stories/2004/jun04/ground.htm.

Gerber, Elisabeth R. 1996. Legislative Response to the Threat of Popular Initiatives. *American Journal of Political Science* 40:99–128.

Gerber, Elisabeth R., and Justin H. Phillips. 2005. Evaluating the Effects of Direct Democracy on Public Policy: California's Urban Growth Boundaries. *American Politics Research* 33:310–330.

Gerber, Elisabeth R., and Justin H. Phillips. 2004. Direct Democracy and Land Use Policy: Exchanging Public Goods for Development Rights. *Urban Studies* 41:463–479.

Gerber, Elisabeth R., and Justin H. Phillips. 2003. Development Ballot Measures, Interest Group Endorsements, and the Political Geography of Growth Preferences. *American Journal of Political Science* 47:625–639.

Gerth, H. H., and C. Wright Mills, eds. 1946. *From Max Weber, Essays in Sociology*. New York: Oxford University Press.

Gertner, Jon. 2007. The Future Is Drying Up. *New York Times Magazine*, October 21, 68–77.

Gilens, Martin. 2005. Inequality and Democratic Responsiveness. *Public Opinion Quarterly* 69:778–796.

Gleick, Peter H. 2003. Global Freshwater Resources: Soft-Path Solutions for the 21st Century. *Science* 302:1524–1528.

Gleick, Peter H. 2002. Soft Water Paths. *Nature* 418:373.

Gleick, Peter H., Nicholas L. Cain, Dana Haasz, Christine Henges-Jeck, Catherine Hunt, Michael Kiparsky, Marcus Moench, Meena Palaniappan, Veena Srinivasan, and Gary H. Wolff. 2004. *The World's Water 2004–2005: The Biennial Report on Freshwater Resources*. Washington, D.C.: Island Press.

Gleick, Peter H., Heather Cooley, David Katz, Emily Lee, Jason Morrison, Meena Palaniappan, Andrea Samulon, and Gary H. Wolff. 2006. *The World's Water 2006–2007: The Biennial Report on Freshwater Resources*. Washington, D.C.: Island Press.

Gleick, Peter H., Gary Wolff, Elizabeth L. Chalecki, and Rachel Reyes. 2002. *New Economy of Water: The Risks and Benefits of Globalization and Privatization of Fresh Water*. Oakland, Calif.: Pacific Institute.

Glennon, Robert. 2002. *Water Follies: Groundwater Pumping and the Fate of America's Fresh Waters*. Washington, D.C.: Island Press.

Global Water Partnership. 2002. *Towards Water Security: A Framework for Action*. Stockholm: Global Water Partnership.

Godwin, R. Kenneth, Helen M. Ingram, and Dean E. Mann. 1985. Introduction: Water Resources and Public Policy. *Review of Policy Research* 5:349–352.

Gomez, Santos V., and Arlene K. Wong. 1997. *Our Water, Our Future: The Need for New Voices in California Water Policy*. Oakland, Calif.: Pacific Institute.

Gottlieb, Robert, and Margaret FitzSimmons. 1991. *Thirst for Growth: Water Agencies as Hidden Government in California*. Tucson: University of Arizona Press.

Governor's Center for Local Government Services. 2002. *Municipal Authorities in Pennsylvania*. Harrisburg: Pennsylvania Department of Community and Economic Development.

Greene, William H. 2003. *Econometric Analysis*. 5th ed. Upper Saddle River, N.J.: Prentice Hall.

Haarmeyer, David, and Debra G. Coy. 2002. An Overview of Private Sector Participation in the Global and US Water and Wastewater Sector. In *Reinventing Water and Wastewater Systems: Global Lessons for Improving Water Management*, ed. Paul Seidenstat, David Haarmeyer, and Simon Hakim, 7–28. New York: John Wiley.

Haeseler, Rob. 1995. Big East Bay Water Fight May Dry Up. *San Francisco Chronicle*, June 26, A1.

Hajnal, Zoltan L., and Paul G. Lewis. 2003. Municipal Institutions and Voter Turnout in Local Elections. *Urban Affairs Review* 38:645–668.

Hall, Darwin C. 2000. Public Choice and Water Rate Design. In *The Political Economy of Water Pricing Reforms*, ed. Aniel Dinar, 189–212. New York: Oxford University Press for the World Bank.

Halligan Project Must Remain a Priority (editorial). 2005. *Fort Collins Coloradoan*, July 10, B6.

Hallissy, Erin. 1997. No EBMUD Water for Big Development. *San Francisco Chronicle*, April 14, A13.

Halstuk, Martin. 1990. EBMUD Race Centers on Buckhorn Reservoir. *San Francisco Chronicle*, October 31, 2.

Hamilton, David K. 2000. Organizing Government Structure and Governance Functions in Metropolitan Areas in Response to Growth and Change: A Critical Overview. *Journal of Urban Affairs* 22:65–84.

Hanak, Ellen. 2008. Is Water Policy Limiting Residential Growth? Evidence from California. *Land Economics* 84:31–50.

Hanak, Ellen. 2005. *Water for Growth: California's New Frontier*. San Francisco: Public Policy Institute of California.

Hanak, Ellen, and Margaret K. Browne. 2006. Linking Housing Growth to Water Supply. *Journal of the American Planning Association* 72:154–166.

Hanak, Ellen, and Ada Chen. 2007. Wet Growth: Effects of Water Policies on Land Use in the American West. *Journal of Regional Science* 47:85–108.

Hanak, Ellen, and Antonina Simeti. 2004. *Water Supply and Growth in California: A Survey of City and County Land-Use Planners*. San Francisco: Public Policy Institute of California.

Harris, Bernard. 2003. Candidates Comment on Growth and Development. *Lancaster New Era*, April 28, B1.

Hawkins, Robert B., Jr. 1976. *Self Government by District: Myth and Reality*. Stanford, Calif.: Hoover Institution Press.

Heikkila, Tanya, and Andrea Gerlak. 2005. The Formation of Large-Scale Collaborative Resource Management Institutions: Clarifying the Roles of Stakeholders, Science, and Institutions. *Policy Studies Journal* 33:583–612.

Heikkila, Tanya, and Kimberley Roussin Isett. 2007. Citizen Involvement and Performance Management in Special-Purpose Governments. *Public Administration Review* 67:238–248.

Henriques, Diana B. 1986. *The Machinery of Greed: Public Authority Abuse and What to Do about It*. Lexington, Mass.: Lexington Books.

Herman, Dennis J. 1992. Sometimes There's Nothing Left to Give: The Justification for Denying Water Service to New Consumers to Control Growth. *Stanford Law Review* 44:429–470.

Hernon, Brian. 2003a. Developers Face Supply/Demand Dilemma in E. Cocalico. *Lancaster Intelligencer Journal*, January 22, B6.

Hernon, Brian. 2003b. E. Cocalico Officials Host Hearing on Regional Comprehensive Plan. *Lancaster Intelligencer Journal*, April 30, B6.

Hernon, Brian. 2003c. E. Cocalico Planners Don't Back Rezoning Plan. *Lancaster Intelligencer Journal*, July 30, B1.

Hernon, Brian. 2003d. Water Supply Problem Persists in East Cocalico Twp. *Lancaster Intelligencer Journal*, October 8, B5.

Hewitt, Julie A. 2000. An Investigation into the Reasons Why Water Utilities Choose Particular Residential Rate Structures. In *The Political Economy of Water Pricing Reforms*, ed. Aniel Dinar, 259–278. New York: Oxford University Press for the World Bank.

Hewitt, Julie A., and W. Michael Hanemann. 1995. A Discrete/Continuous Choice Approach to Residential Water Demand under Block Rate Pricing. *Land Economics* 71:173–192.

Hickey, Harry E. 2002. *Fire Suppression Rating Schedule Handbook 2002*. Louisville, Ky.: Chicago Spectrum.

Holst, Arthur. 2007. The Philadelphia Water Department and the Burden of History. *Public Works Management and Policy* 11:233–238.

Hooghe, Liesbet, and Gary Marks. 2003. Unraveling the Central State, but How? Types of Multi-level Governance. *American Political Science Review* 97:233–243.

Hundley, Norris, Jr. 2001. *The Great Thirst: Californians and Water, a History*. Rev. ed. Berkeley and Los Angeles: University of California Press.

Hutson, Susan S., Nancy L. Barber, Joan F. Kenny, Kristin S. Linsey, Deborah S. Lumia, and Molly A. Maupin. 2004. *Estimated Use of Water in the United States in 2000.* Circular 1268. Denver: U.S. Geological Survey.

Hytha, Michael. 1998. Lack of Water May Kill Tassajara Preserve Plans. *San Francisco Chronicle,* September 18, A19.

Hytha, Michael. 1997. Supervisors Rip Impact Report on Mt. Diablo Plans. *San Francisco Chronicle,* July 8, A11.

Ihlanfeldt, Keith R., and Timothy M. Shaughnessy. 2004. An Empirical Investigation of the Effects of Impact Fees on Housing and Land Use Markets. *Regional Science and Urban Economics* 34:639–661.

Ingram, Helen. 1990. *Water Politics: Continuity and Change.* Albuquerque: University of New Mexico Press.

Jacobus, Patricia, and Michael Hytha. 1998. How Developers' Dream Died in Contra Costa. *San Francisco Chronicle,* May 26, A1.

Jehl, Douglas. 2003. As Cities Move to Privatize Water, Atlanta Steps Back. *New York Times,* February 10, A1.

Jehl, Douglas. 2002. Development and a Drought Cut Carolinas' Water Supply. *New York Times,* August 29, A1.

Jeong, Moon-Gi. 2006. Local Choices for Development Impact Fees. *Urban Affairs Review* 41:338–357.

Jeong, Moon-Gi, and Richard C. Feiock. 2006. Impact Fees, Growth Management, and Development: A Contractual Approach to Local Policy and Governance. *Urban Affairs Review* 41:749–768.

Johnson, Jason B. 2002. Pushing the Limits: Contra Costa Residents Say Growth out of Control. *San Francisco Chronicle,* June 21, A17.

Jones, Bill. 1994. *Statement of Vote: November 8, 1994 General Election.* Sacramento: California Secretary of State.

Jones, Bryan D. 1981. Party and Bureaucracy: The Influence of Intermediary Groups on Urban Public Service Delivery. *American Political Science Review* 75:688–700.

Jones, Bryan D. 1973. Competitiveness, Role Orientations, and Legislative Responsiveness. *Journal of Politics* 35:924–947.

Jones, Bryan D., and Frank R. Baumgartner. 2005. *The Politics of Attention: How Government Prioritizes Problems.* Chicago: University of Chicago Press.

Jones, Victor. 1966. Metropolitan Authorities. In *Metropolitan Politics: A Reader,* ed. Michael N. Danielson, 238–245. Boston: Little, Brown.

Jones, Victor. 1942. *Metropolitan Government.* Chicago: University of Chicago Press.

Kahrl, William. 1982. *Water and Power.* Berkeley and Los Angeles: University of California Press.

Kaiser, Edward J., Raymond J. Burby, and David H. Moreau. 1988. Local Governments' Use of Water and Sewer Impact Fees and Related Policies: Current Practice in the Southeast. In *Development Impact Fees: Policy Rationale, Practice, Theory, and Issues*, ed. Arthur C. Nelson, 22–36. Chicago: Planners Press.

Karnig, Albert K. 1975. "Private-Regarding" Policy, Civil Rights Groups, and the Mediating Impact of Municipal Reforms. *American Journal of Political Science* 19:91–106.

Karnig, Albert K., and Susan Welch. 1982. Electoral Structure and Black Representation on City Councils. *Social Science Quarterly* 63:153–161.

Kasler, Dale. 2002. Builders Developing Big Thirst for Water. *Sacramento Bee*, September 23, A1.

Katz, Bruce, ed. 2000. *Reflections on Regionalism*. Washington, D.C.: Brookings Institution.

King, Dwight. 2001. County Board Ready for New Chairman. *Tri-Valley Herald*, December 24. Retrieved from LexisNexis Academic Universe database.

Kingdon, John W. 1973. *Congressmen's Voting Decisions*. New York: Harper and Row.

Koehler, David H., and Margaret T. Wrightson. 1987. Inequality in the Delivery of Urban Services: A Reconsideration of the Chicago Parks. *Journal of Politics* 49:80–99.

Kundzewicz, Zbigniew W., Luis José Mata, Nigel W. Arnell, Petra Döll, Pavel Kabat, Blanca Jiménez, Kathleen A. Miller, Taikan Oki, Zekai Sen, and Igor A. Shiklomanov. 2007. Freshwater Resources and Their Management. In *Climate Change 2007: Impacts, Adaptation, and Vulnerability. Contribution of Working Group II to the Fourth Assessment Report of the Intergovernmental Panel on Climate Change*, ed. Martin L. Parry, Osvaldo F. Canziani, Jean P. Palutikof, Paul J. van der Linden, and Clair E. Hanson, 173–210. Cambridge, U.K.: Cambridge University Press.

Lach, Denise, Helen Ingram, and Steve Rayner. 2005. Maintaining the Status Quo: How Institutional Norms and Practices Create Conservative Water Organizations. *Texas Law Review* 83:2027–2053.

Lancaster County Planning Commission. 2004. *Cocalico Region Strategic Comprehensive Plan*. Lancaster, Pa.: Lancaster County Planning Commission, November 2004.

Landis, John, Michael Larice, Deva Dawson, and Lan Deng. 1999. *Pay to Play: Residential Development Fees in California Cities and Counties, 1999*. Sacramento: California Department of Housing and Community Development.

Leigland, James. 1994. Public Authorities and the Determinants of Their Use by State and Local Governments. *Journal of Public Administration Research and Theory* 4:521–544.

LeRoux, Kelly, and Jered B. Carr. 2007. Explaining Local Government Cooperation on Public Works: Evidence from Michigan. *Public Works Management and Policy* 12:344–358.

Levine, Daniel S. 2000. Debate Rages over Role of Water Agencies. *San Francisco Business Times*, July 28, 1.

Levy, Frank S., Arnold J. Meltsner, and Aaron B. Wildavsky. 1974. *Urban Outcomes: Schools, Streets, and Libraries*. Berkeley and Los Angeles: University of California Press.

Lineberry, Robert L. 1977. *Equality and Urban Policy: The Distribution of Municipal Public Services*. Beverly Hills, Calif.: Sage.

Lineberry, Robert L., and Edmund P. Fowler. 1967. Reformism and Public Policies in American Cities. *American Political Science Review* 61:701–716.

Liner, Gaines H. 1990. Annexation Rates and Institutional Constraints. *Growth and Change* 21:80–94.

Liner, Gaines H., and Rob Roy McGregor. 1996. Institutions and the Market for Annexable Land. *Growth and Change* 27:55–74.

Little Hoover Commission. 2000. *Special Districts: Relics of the Past or Resources for the Future?* Report no. 155. Sacramento: Little Hoover Commission.

Locke, Michelle. 1993. Development Hits Rising Tide of Water and Traffic Troubles. *Los Angeles Times*, October 31, A1.

Logan, John R., and Harvey L. Molotch. 1987. *Urban Fortunes: The Political Economy of Place*. Berkeley and Los Angeles: University of California Press.

Lowi, Theodore J. 1969. *The End of Liberalism: Ideology, Policy, and the Crisis of Public Authority*. New York: Norton Press.

Lubell, Mark, Richard C. Feiock, and Edgar Ramirez. 2005. Political Institutions and Conservation by Local Governments. *Urban Affairs Review* 40:706–729.

Lubell, Mark, Mark Schneider, John T. Scholz, and Mihriye Mete. 2002. Watershed Partnerships and the Emergence of Collective Action Institutions. *American Journal of Political Science* 46:148–163.

Luoma, Jon R. 2002. Water for Profit: The Price of Water. *Mother Jones* (November–December): 34–37.

Macauley, David. 2007. Currituck Lifts Halt on Water Hook-Ups. *Daily Advance*, February 2, A1.

MacManus, Susan A. 1981. Special District Governments: A Note on Their Use as Property Tax Relief Mechanisms in the 1970s. *Journal of Politics* 43:1206–1214.

MacManus, Susan, and Robert Thomas. 1979. Expanding the Tax Base: Does Annexation Make a Difference? *The Urban Interest* 1:15–28.

Manochio, Matt. 2005. Roxbury to Extend Limits on Its Water. *Roxbury Daily Record*, December 10, 1.

Martin, William E., Helen M. Ingram, Nancy K. Laney, and Adrian H. Griffin. 1984. *Saving Water in a Desert City*. Washington, D.C.: Resources for the Future.

Maryland Department of the Environment. 2003. *Guidance for Maryland Public Water Systems on Best Management Practices for Improving Water Conservation and Water Use Efficiency*. Baltimore: Maryland Department of the Environment.

Mayer, Christopher J., and C. Tsuriel Somerville. 2000. Land Use Regulation and New Construction. *Regional Science and Urban Economics* 30:639–662.

Mazmanian, Daniel A., and Michael E. Kraft, eds. 1999. *Toward Sustainable Communities: Transition and Transformations in Environmental Policy*. Cambridge, Mass.: MIT Press.

McCabe, Barbara C. 2000. Special District Formation among the States. *State and Local Government Review* 32:121–131.

McCandlish, Laura. 2006. Carroll City's Growth Halted: State Officials Order Development Frozen in Westminster until New Sources of Water Found. *Baltimore Sun*, September 28, A1.

McCubbins, Mathew D., Roger G. Noll, and Barry R. Weingast. 1989. Structure and Process, Politics and Policy: Administrative Arrangements and the Political Control of Agencies. *Virginia Law Review* 75:431–482.

McCubbins, Mathew D., Roger G. Noll, and Barry R. Weingast. 1987. Administrative Procedures as Instruments of Political Control. *Journal of Law, Economics, and Organization* 3:243–277.

McCubbins, Mathew D., and Thomas Schwartz. 1984. Congressional Oversight Overlooked: Police Patrols versus Fire Alarms. *American Journal of Political Science* 28:165–179.

McGinnis, Michael D., ed. 1999. *Polycentricity and Local Public Economies: Readings from the Workshop in Political Theory and Policy Analysis*. Ann Arbor: University of Michigan Press.

Mehay, Stephen L. 1984. The Effect of Governmental Structure on Special District Expenditures. *Public Choice* 44:339–348.

Meier, Kenneth J., Eric Gonzales Juenke, Robert D. Wrinkle, and J. L. Polinard. 2005. Structural Choices and Representational Biases: The Post-election Color of Representation. *American Journal of Political Science* 49:758–768.

Meier, Kenneth J., and Thomas M. Holbrook. 1992. I Seen My Opportunities and I Took 'Em: Political Corruption in the American States. *Journal of Politics* 54:135–155.

Meltzer, Allan H., and Scott F. Richard. 1981. A Rational Theory of the Size of Government. *Journal of Political Economy* 89:214–227.

Miller, Alex. 1992. Water Supply, Growth Needs May Not Jibe. *The Independent*, February 12, 1.

Miller, Gary J. 1981. *Cities by Contract: The Politics of Municipal Incorporation*. Cambridge, Mass.: MIT Press.

Miller, Inga. 2002. Alamo Creek Likely to Dominate Water Board Race. *Oakland Tribune*, August 15, A1.

Miller, Warren E., and Donald E. Stokes. 1963. Constituency Influence in Congress. *American Political Science Review* 57:45–56.

Minge, David. 1976. Special Districts and the Level of Public Expenditures. *Journal of Urban Law* 53:701–718.

Mitchell, David L., and W. Michael Hanemann. 1994. *Setting Urban Water Rates for Efficiency and Conservation: A Discussion of Issues*. Sacramento: California Urban Water Conservation Council.

Mitchell, Jerry. 1990. The Policy Activities of Public Authorities. *Policy Studies Journal* 18:928–942.

Mladenka, Kenneth R. 1989. The Distribution of an Urban Service: The Changing Role of Race and Politics. *Urban Affairs Quarterly* 24:556–583.

Mladenka, Kenneth R. 1980. The Urban Bureaucracy and the Chicago Political Machine: Who Gets What and the Limits to Political Control. *American Political Science Review* 74:991–998.

Moe, Terry M. 1982. Regulatory Performance and Presidential Administration. *American Journal of Political Science* 26:197–224.

Molotch, Harvey L. 1976. The City as a Growth Machine: Toward a Political Economy of Place. *American Journal of Sociology* 82:309–330.

Morgan, David R., and Michael W. Hirlinger. 1991. Intergovernmental Service Contracts: A Multivariate Explanation. *Urban Affairs Quarterly* 27:128–144.

Morgan, David R., and John P. Pelissero. 1980. Urban Policy: Does Political Structure Matter? *American Political Science Review* 74:999–1006.

Mosely, Brian. 2006. Bell Buckle Water System Can Handle Growth. *Shelbyville Times-Gazette*, December 14. Retrieved from Access World News database.

Mr. Suozzi's Next Crusade (editorial). 2007. *New York Times*, March 18, Long Island sec., 15.

Mullin, Megan. 2008. The Conditional Effect of Specialized Governance on Public Policy. *American Journal of Political Science* 52:124–140.

Mullin, Megan. 2009. Special Districts versus Contracts: Complements or Substitutes? In *Self-Organizing Federalism: Collaborative Mechanisms to Mitigate Institutional Collective Action*, ed. Richard C. Feiock and John T. Scholz. New York: Cambridge University Press.

Mullin, Megan. 2007. California Water: A Case Study in Federalism. In *Governing California: Politics, Government, and Public Policy in the Golden State*, 2d ed., ed. Gerald C. Lubenow, 213–234. Berkeley, Calif.: Institute of Governmental Studies Press.

Mullin, Megan, Gillian Peele, and Bruce E. Cain. 2004. City Caesars? Institutional Structure and Mayoral Success in Three California Cities. *Urban Affairs Review* 40:19–43.

Natale, Darlene White. 1997a. Accepting West View Offer of Water Becomes Moot. *Pittsburgh Post-Gazette*, March 26, N4.

Natale, Darlene White. 1997b. Company Hired to Help Create New Zoning Plan. *Pittsburgh Post-Gazette,* November 26, N5.

Natale, Darlene. 1996. Water Service a Fluid Question. *Pittsburgh Post-Gazette,* October 20, NW2.

National Assessment Synthesis Team, U.S. Global Change Research Program. 2001. *Climate Change Impacts on the United States.* New York: Cambridge University Press.

National Research Council, Water Science and Technology Board. 2004. *Confronting the Nation's Water Problems: The Role of Research.* Washington, D.C.: National Academy Press.

National Research Council, Water Science and Technology Board. 2002a. *Estimating Water Use in the United States: A New Paradigm for the National Water-Use Information Program.* Washington, D.C.: National Academy Press.

National Research Council, Water Science and Technology Board. 2002b. *Privatization of Water Services in the United States: An Assessment of Issues and Experience.* Washington, D.C.: National Academy Press.

National Research Council, Water Science and Technology Board. 2001. *Envisioning the Agenda for Water Resources Research in the Twenty-First Century.* Washington, D.C.: National Academy Press.

Nealon, Patricia. 1991. Brockton Lifts Ban on New Hookups to Water Supply. *Boston Globe,* June 23, Metro/Region sec., 24.

Needham, Jerry. 2007a. Brackish Water Could Ease San Antonio's Drought Pains. *San Antonio Express-News,* May 25, A1.

Needham, Jerry. 2007b. Stored Water Is Focus of Fight. *San Antonio Express-News,* August 8, A1.

Needham, Jerry. 2006a. BexarMet Loses Court Case over Expansion of Services. *San Antonio Express-News,* April 19, B2.

Needham, Jerry. 2006b. BexarMet to Sell 5 Systems to SAWS. *San Antonio Express-News,* September 6, B3.

Needham, Jerry. 2005a. BexarMet to Try to Tap into Trinity. *San Antonio Express-News,* July 28, B8.

Needham, Jerry. 2005b. New Bill Would Rein in BexarMet. *San Antonio Express-News,* April 22, 3B.

Needham, Jerry. 2005c. Water War is Boiling over Service in Northern Bexar. *San Antonio Express-News,* August 20, A1.

New York State Comptroller's Office. 2007. *Town Special Districts in New York: Background, Trends, and Issues.* Albany, N.Y.: Office of the State Comptroller.

North, Douglass C. 1990. *Institutions, Institutional Change, and Economic Performance.* Cambridge, U.K.: Cambridge University Press.

Oates, Wallace E. 1989. Searching for Leviathan: A Reply and Some Further Reflections. *American Economic Review* 79:578–583.

Oliver, J. Eric, and Shang Ha. 2007. Vote Choice in Suburban Elections. *American Political Science Review* 101:393–408.

Olmstead, Sheila M., W. Michael Hanemann, and Robert N. Stavins. 2005. *Do Consumers React to the Shape of Supply? Water Demand under Heterogeneous Price Structures*. Regulatory Policy Program Working Paper RPP-2005-05. Cambridge, Mass.: Center for Business and Government, John F. Kennedy School of Government, Harvard University.

Olson, Mancur. 1965. *The Logic of Collective Action*. Cambridge, Mass.: Harvard University Press.

Orfield, Myron. 1997. *Metropolitics: A Regional Agenda for Community and Stability*. Washington, D.C., and Cambridge, Mass.: Brookings Institution and Lincoln Institute of Land Policy.

Organization for Economic Cooperation and Development (OECD). 2003. *Social Issues in the Provision and Pricing of Water Services*. Paris: OECD.

Organization for Economic Cooperation and Development (OECD). 1999. *Household Water Pricing in OECD Countries*. Paris: OECD.

Ostrom, Elinor. 2005. *Understanding Institutional Diversity*. Princeton, N.J.: Princeton University Press.

Ostrom, Elinor. 1990. *Governing the Commons*. New York: Cambridge University Press.

Ostrom, Elinor. 1972. Metropolitan Reform: Propositions Derived from Two Traditions. *Social Science Quarterly* 53:474–493.

Ostrom, Vincent. 1953. *Water and Politics: A Study of Water Policies and Administration in the Development of Los Angeles*. Los Angeles: Haynes Foundation.

Ostrom, Vincent, Robert Bish, and Elinor Ostrom. 1988. *Local Government in the United States*. San Francisco: ICS Press.

Ostrom, Vincent, Charles M. Tiebout, and Robert Warren. 1961. The Organization of Government in Metropolitan Areas: A Theoretical Inquiry. *American Political Science Review* 55:831–842.

Pagano, Michael A. 1999. Metropolitan Limits: Intrametropolitan Disparities and Governance in U.S. Laboratories of Democracy. In *Governance and Opportunity in Metropolitan America*, ed. Alan Altshuler, William Morrill, Harold Wolman, and Faith Mitchell, 253–295. Washington, D.C.: National Academy Press.

Page, Benjamin I., and Robert Y. Shapiro. 1983. Effects of Public Opinion on Policy. *American Political Science Review* 77:175–190.

Page, G. William, and Lawrence Susskind. 2007. Five Important Themes in the Special Issue on Planning for Water. *Journal of the American Planning Association* 73:141–145.

Pelissero, John P., and Timothy B. Krebs. 1997. City Council Legislative Committees and Policy-making in Large United States Cities. *American Journal of Political Science* 41:499–518.

Perrenod, Virginia Marion. 1984. *Special Districts, Special Purposes: Fringe Governments and Urban Problems in the Houston Area.* College Station: Texas A&M University Press.

Peterson, Paul E. 1981. *City Limits.* Chicago: University of Chicago Press.

Pint, Ellen M. 1999. Household Responses to Increased Water Rates during the California Drought. *Land Economics* 75:246–247.

Pisani, Donald. 2002. *Water and American Government: The Reclamation Bureau, National Water Policy, and the West, 1902–1935.* Berkeley: University of California Press.

Porter, Douglas R. 1987. Exactions and the Development Process. In *Development Exactions*, ed. James E. Frank and Robert M. Rhodes, 104–122. Chicago: Planners Press.

Porter, Douglas R., Ben C. Lin, and Richard B. Peiser. 1987. *Special Districts: A Useful Technique for Financing Infrastructure.* Washington, D.C.: Urban Land Institute.

Portney, Kent E. 2003. *Taking Sustainable Cities Seriously: Economic Development, the Environment, and Quality of Life in American Cities.* Cambridge, Mass.: MIT Press.

Post, Stephanie Shirley. 2002. Local Government Cooperation: The Relationship Between Metropolitan Area Government Geography and Service Provision. Paper presented at the annual meeting of the American Political Science Association, August 29–September 1, Boston.

Purdum, Elizabeth D., and James E. Frank. 1987. Community Use of Exactions: Results of a National Survey. In *Development Exactions*, ed. James E. Frank and Robert M. Rhodes, 123–152. Chicago: Planners Press.

Reisner, Marc. 1986. *Cadillac Desert: The American West and Its Disappearing Water.* New York: Penguin Books.

Reiterman, Tim. 2006. Small Towns Tell a Cautionary Tale about the Private Control of Water. *Los Angeles Times*, May 30, A1.

Renwick, Mary E., and Sandra O. Archibald. 1998. Demand Side Management Policies for Residential Water Use: Who Bears the Conservation Burden? *Land Economics* 74:343–359.

Rosenthal, Alan. 1990. *Governors & Legislators: Contending Powers.* Washington, D.C.: CQ Press.

Rusk, David. 1993. *Cities Without Suburbs.* Washington, D.C.: Woodrow Wilson Center Press.

Sabatier, Paul A., Will Focht, Mark Lubell, Zev Trachtenberg, Arnold Vedlitz, and Marty Matlock, eds. 2005. *Swimming Upstream: Collaborative Approaches to Watershed Management.* Cambridge, Mass.: MIT Press.

Saltonstall, David D. 1992. Zone 7 Debates Its Role in Growth. *Tri-Valley Herald*, February 12, A1.

Schattschneider, E. E. 1960. *The Semi-sovereign People*. New York: Holt, Rinehart and Winston.

Schneider, Mark. 1989. *The Competitive City: The Political Economy of Suburbia*. Pittsburgh: University of Pittsburgh Press.

Schneider, Mark, and Paul Teske, with Michael Mintrom. 1995. *Public Entrepreneurs: Agents for Change in American Government*. Princeton, N.J.: Princeton University Press.

Scholz, John T., and Bruce Stiftel, eds. 2005. *Adaptive Governance and Water Conflict: New Institutions for Collaborative Planning*. Washington, D.C.: Resources for the Future.

Schumaker, Paul D., and Russell W. Getter. 1977. Responsiveness Bias in 51 American Communities. *American Journal of Political Science* 21:247–281.

Sharp, Elaine B. 2002. Culture, Institutions, and Urban Officials' Responses to Morality Issues. *Political Research Quarterly* 55:861–883.

Sharp, Elaine B. 1986. The Politics and Economics of the New City Debt. *American Political Science Review* 80:1271–1288.

Shepsle, Kenneth. 1979. Institutional Arrangements and Equilibrium in Multidimensional Voting Models. *American Journal of Political Science* 23:27–59.

Shepsle, Kenneth, and Barry Weingast. 1987. The Institutional Foundations of Committee Power. *American Political Science Review* 81:85–104.

Shigley, Paul, and John Krist. 2002. Drip Drip Drip: A National Water Shortage Leads to Some Tough Planning Measures. *Planning* 68:4–8.

Singell, Larry D., and Jane H. Lillydahl. 1990. An Empirical Examination of the Effect of Impact Fees on the Housing Market. *Land Economics* 66:82–92.

Smith, Robert G. 1964. *Public Authorities, Special Districts, and Local Government*. Washington, D.C.: National Association of Counties Research Foundation.

Snitow, Alan, and Deborah Kaufman, with Michael Fox. 2007. *Thirst: Fighting the Corporate Theft of Our Water*. San Francisco: Jossey-Bass and John Wiley.

Solley, Wayne B., Robert R. Pierce, and Howard A. Perlman. 1998. *Estimated Use of Water in the United States in 1995*. Circular 1200. Denver: U.S. Geological Survey.

Speir, Cameron, and Kurt Stephenson. 2002. Does Sprawl Cost Us All? Isolating the Effects of Housing Patterns on Public Water and Sewer Costs. *Journal of the American Planning Association* 68:56–70.

Spulber, Nicolas, and Asghar Sabbaghi. 1998. *Economics of Water Resources: From Regulation to Privatization*. 2d ed. Boston: Kluwer Academic.

Stein, Robert M. 1990. *Urban Alternatives: Public and Private Markets in the Provision of Local Services*. Pittsburgh: University of Pittsburgh Press.

Stimson, James A. 1985. Regression in Space and Time: A Statistical Essay. *American Journal of Political Science* 29:914–947.

Stone, Clarence N. 1980. Systemic Power in Community Decision Making: A Restatement of Stratification Theory. *American Political Science Review* 74:978–990.

Sweeney, James W. 1999. Concern over RP Land-Use Plan. *Santa Rosa Press-Democrat*, November 10, B1.

Teodoro, Manuel P. 2008. Contingent Professionalism: Bureaucratic Mobility and the Adoption of Water Conservation Rates. Paper presented at the annual meeting of the Midwest Political Science Association, April 3–6, Chicago.

Texas Water Development Board. 2006. *2006 South Central Texas Regional Water Plan*. Austin: Texas Water Development Board.

Thomas, Robert D., David Hawes, and Bill Calderon. 2003. Creating "Federal Bias" through the Use of Inside-the-City Special Districts: The Case of Houston. Paper presented at the annual meeting of the Midwest Political Science Association, April 3–6, Chicago.

Thomas, Robert D., and Richard W. Murray. 1991. *Progrowth Politics: Change and Governance in Houston*. Berkeley, Calif.: IGS Press.

Tiebout, Charles M. 1956. A Pure Theory of Local Expenditures. *Journal of Political Economy* 64:416–424.

Timmins, Christopher. 2002. Does the Median Voter Consume Too Much Water? Analyzing the Redistributive Role of Residential Water Bills. *National Tax Journal* 55:687–702.

Umble, Chad. 2003. East Cocalico Officials Approve Updated Zoning Map. *Lancaster Intelligencer Journal*, December 10, B6-A.

United Nations Conference on Environment and Development. 1992. *Agenda 21: Report of the United Nations Conference on Environment and Development*. New York: United Nations.

United Nations World Water Assessment Program. 2003. *Water for People, Water for Life: First UN World Water Development Report*. Paris and Oxford: United Nations Educational, Scientific, and Cultural Organization and Berghahn Books.

U.S. Census Bureau. 2005a. *2002 Census of Governments: State and Local Government Finances File* [Computer file]. Washington, D.C.: U.S. Department of Commerce.

U.S. Census Bureau. 2005b. *2002 Census of Governments*. Vol. 1, no. 2: *Individual State Descriptions*. Washington, D.C.: U.S. Department of Commerce.

U.S. Census Bureau. 2005c. *2002 Census of Governments*. Vol. 4, no. 5: *Compendium of Government Finances*. Washington, D.C.: U.S. Department of Commerce.

U.S. Census Bureau. 2004. *2002 Census of Governments: Local Government Employment and Payroll Data* [Computer file]. Washington, D.C.: U.S. Department of Commerce.

U.S. Census Bureau. 2002a. *2002 Census of Governments: Government Organization Public Use Files* [Computer file]. Washington, D.C.: U.S. Department of Commerce.

U.S. Census Bureau. 2002b. *2002 Census of Governments*. Vol. 1, no. 1: *Government Organization.* Washington, D.C.: U.S. Department of Commerce.

U.S. Census Bureau. 1999. *1997 Census of Governments*. Vol. 1, no. 1: *Government Organization.* Washington, D.C.: U.S. Department of Commerce.

U.S. Census Bureau. 1995. *1992 Census of Governments*. Vol. 1, no. 2: *Popularly Elected Officials.* Washington, D.C.: U.S. Department of Commerce.

U.S. Census Bureau. 1994. *1992 Census of Governments: Government Organization Public Use Files* [Computer file]. Washington, D.C.: U.S. Department of Commerce.

U.S. Census Bureau. 1990. *1987 Census of Governments: Government Organization File* [Computer file]. Washington, D.C.: U.S. Department of Commerce [producer]. Ann Arbor, Mi.: Inter-university Consortium for Political and Social Research [distributor].

U.S. Conference of Mayors Urban Water Council. 1997. *A Status Report on Public/Private Partnerships in Municipal Water and Wastewater Systems: A 261 City Survey.* Washington, D.C.: Urban Water Council.

U.S. Environmental Protection Agency (EPA). 2006. *Growing toward More Efficient Water Use: Linking Development, Infrastructure, and Drinking Water Policies.* Development, Community, and Environment Division, EPA 230-R-06-001. Washington, D.C.: U.S. EPA.

U.S. Environmental Protection Agency (EPA). 2005. *Drinking Water Infrastructure Needs Survey and Assessment.* Third Report to Congress. Office of Water, EPA 816-R-05-001. Washington, D.C.: U.S. EPA.

U.S. Environmental Protection Agency (EPA). 2002a. *Cases in Water Conservation: How Efficiency Programs Help Water Utilities Save Water and Avoid Costs.* Office of Water, EPA 832-B-02-003. Washington, D.C.: U.S. EPA.

U.S. Environmental Protection Agency (EPA). 2002b. *The Clean Water and Drinking Water Infrastructure Gap Analysis.* Office of Water, EPA 816-R-02-020. Washington, D.C.: U.S. EPA.

U.S. Environmental Protection Agency (EPA). 2002c. *Community Water System Survey 2000.* Vol. 1: *Overview.* Office of Water, EPA 815-R-02-005A. Washington, D.C.: U.S. EPA.

U.S. Environmental Protection Agency (EPA). 1997a. *Community Water System Survey.* Vol. 1: *Overview.* Office of Water, EPA 815-R-97-001A. Washington, D.C.: U.S. EPA.

U.S. Environmental Protection Agency (EPA). 1997b. *Community Water System Survey*. Vol. 2: *Detailed Survey Result Tables and Methodology Report*. Office of Water, EPA 815-R-97-001B. Washington, D.C.: U.S. EPA.

U.S. General Accounting Office (GAO). 2000. *Survey of Local Growth Issues*. Report RCED-00-272. Washington, D.C.: U.S. Government Printing Office.

Vonderbrueggen, Lisa. 1998. Livermore Files Suit over Zone 7 Water. *Valley Times*, April 2, A3.

Vonderbrueggen, Lisa. 1996. DSRSD May Study Dropping Zone 7. *Valley Times*, October 18, A3.

Vonheeder, Georgean, and G. T. McCormick. 1992. Administering Policy (Readers' Forum). *Valley Times*, February 18, A8.

Wagner, Richard E., and Warren E. Weber. 1975. Competition, Monopoly, and the Organization of Government in Metropolitan Areas. *Journal of Law and Economics* 18:661–684.

Walsh, Annmarie H. 1978. *The Public's Business: The Politics and Practices of Government Corporations*. Cambridge, Mass.: MIT Press.

Weikel, Dan. 2008. Sewage in O.C. Goes Full Circle: Intensive Cleaning Will Yield Drinking Water and a Buffer Against Import Cost Hikes and Shortages. *Los Angeles Times*, January 2, A1.

Weingast, Barry R., and Mark J. Moran. 1983. Bureaucratic Discretion or Congressional Control? Regulatory Policymaking by the Federal Trade Commission. *Journal of Political Economy* 91:765–800.

Weiskind, Ida. 1995a. Housing Development Water Is Being Sought. *Pittsburgh Post-Gazette*, July 26, N4.

Weiskind, Ida. 1995b. Residents Lobby to Delay Signing of Water Contract. *Pittsburgh Post-Gazette*, October 11, N5.

Weiskind, Ida. 1995c. Zoning Delays Water Deal for Adams. *Pittsburgh Post-Gazette*, November 26, NW3.

Welch, Susan. 1990. The Impact of At-Large Elections on the Representation of Blacks and Hispanics. *Journal of Politics* 52:1050–1076.

Welch, Susan, and Timothy Bledsoe. 1988. *Urban Reform and Its Consequences: A Study in Representation*. Chicago: University of Chicago Press.

Werkman, Janet, and David L. Westerling. 2000. Privatizing Municipal Water and Wastewater Systems. *Public Works Management and Policy* 5:52–68.

Weschler, Louis F., Alvin H. Mushkatel, and James E. Frank. 1987. Politics and Administration of Development Exactions. In *Development Exactions*, ed. James E. Frank and Robert M. Rhodes, 15–41. Chicago: Planners Press.

Wescoat, James L., and Gilbert F. White. 2003. *Water for Life*. Cambridge, U.K.: Cambridge University Press.

Williamson, Oliver. 1975. *Markets and Hierarchies*. New York: Free Press.

Wilson, James Q. 1980. *The Politics of Regulation*. New York: Basic Books.

Wilson, Woodrow. 1887. The Study of Administration. *Political Science Quarterly* 2:197–222.

Wlezien, Christopher. 2005. The Salience of Political Issues: The Problem with "Most Important Problem." *Electoral Studies* 24:555–579.

Wlezien, Christopher. 2004. Patterns of Representation: Dynamics of Public Preferences and Policy. *Journal of Politics* 66:1–24.

Wolff, Gary, and Eric Hallstein. 2005. *Beyond Privatization: Restructuring Water Systems to Improve Performance.* Oakland, Calif.: Pacific Institute.

Wolfinger, Raymond E., and John Osgood Field. 1966. Political Ethos and the Structure of City Government. *American Political Science Review* 60:306–326.

Wood, Robert C. 1961. *1400 Governments: The Political Economy of the New York Region.* Cambridge, Mass.: Harvard University Press.

Wood, Robert C. 1958. The New Metropolis: Green Belts, Grass Roots, or Gargantua. *American Political Science Review* 52:108–122.

World Health Organization. 2000. *Global Water Supply and Sanitation Assessment 2000 Report.* Geneva: World Health Organization and the United Nations Children's Fund.

World Water Council. 2000. *World Water Vision: Making Water Everybody's Business.* London: Earthscan.

Worster, Donald. 1992. *Rivers of Empire: Water, Aridity, and the Growth of the American West.* New York: Oxford University Press.

Yinger, John. 1998. The Incidence of Development Fees and Special Assessments. *National Tax Journal* 51:23–41.

Zax, Jeffery S. 1989. Is There a Leviathan in Your Neighborhood? *American Economic Review* 79:560–567.

Zone 7 Water Agency. 1997. *Zone 7 Water for Dougherty Valley: Questions and Answers.* Livermore, Ca.: Zone 7 Water Agency.

Index

Note: A page number followed by a *t* indicates a table, *g* a graph, and *m* a map.

Accountability
  institutional design's effect on, 31, 39, 48, 74–79, 183
  metropolitan reform theory on, 31, 34, 38, 39
  of private water companies, 18, 19
  public choice theory on, 8, 34
  of special district officials, 7–8, 31, 34, 36, 39, 43, 75–79, 184
ACIR. *See* Advisory Commission on Intergovernmental Relations
Adams Township dispute, 137–138, 137*m*, 158–160, 172
Adams Township Water Authority, 138, 158–160, 172
Advisory Commission on Intergovernmental Relations (ACIR), 29–30, 112, 209, 216n7
Agenda 21 (international plan), 190
Agreements. *See* Contracts and agreements; Cooperative relationships
Agriculture, 9, 12*g*, 13, 14–15
Air pollution, 15*g*, 16
Alabama, 11, 73, 113
Alameda County, California, 133*m*, 141*t*, 144–146, 147–153, 172
Alameda County Flood Control and Water Conservation District (Zone 7), 141*t*, 147–153, 183
Altshuler, Alan, 82–83, 84

American River, 134
American Water Works Association (AWWA), 51, 63–64, 195, 220n14, 225–226n10
Annexation. *See also* Boundaries; Boundary flexibility
  of Dougherty Valley by DSRSD, 147
  formation of special districts and, 6, 107
  motivations for, 224–225n1
  relationship to cooperative agreements, 107–111, 115–116, 116*t*, 117*g*, 118–121, 120*t*, 122, 184, 225n5, 225n9
  rules on, 111, 112–113, 113*t*
  as supply/demand mismatch solution, 104, 105–106, 111
Antigrowth activists
  benefits of specialized governance to, 151
  challenges for, 49–50
  Dougherty Valley dispute, role in, 132, 133*m*, 142, 150
  East Cocalico dispute, role in, 162
  increased activity, 185–186
  moratoria use on connections, 126
  political conflicts over growth and land use, role in, 185–186
  Rohnert Park and Sonoma County dispute, role in, 165, 168–169
  support of impact fees, 84, 85–86

Antigrowth activists (cont.)
sustainability policies and, 191
Tassajara Valley dispute, role in, 132, 133*m*, 153–154, 156–157
Association of Bay Area Governments survey, 92
Atlanta, Georgia, 18, 19
At-large elections, 76, 78*g*
Auburn Dam, 11
Aurora, Colorado, 176
Avery v. Midland County, 218n20
AWWA. *See* American Water Works Association

Ball v. James, 31, 217–218n20
Berrenda Mesa Water District, 147, 149
Berry, William, 58
Besley, Timothy, 43
Bexar Metropolitan Water District (BexarMet), 109–110, 111, 118, 123–124, 127, 182, 225n8
Bias. *See also* Accountability; Responsiveness
general-purpose venues and, 35, 38
election rules and, 74–79, 77*g*, 78*g*, 80, 183
impact fees use and, 82, 88–91, 94–95, 95*g*, 96*g*, 97–102
increasing block rates use and, 19, 66–67, 69–74, 70*t*, 71*g*, 72*g*, 76, 80
metropolitan reform theory on, 89
public choice theory on, 38, 66, 90
specialized governance and, 19, 38–39, 43–44, 49–51, 66, 89–91, 175
Black & Veatch surveys, 203
Bolinas Public Utility District, 125–126
Bollens, John, 31, 89, 216n7
Bottled water, 215n28
Boundaries. *See also* Annexation
of cities/counties, 28
contiguity between special districts/city or county, 40*t*, 49, 179*t*, 184
contiguity requirement, 113–114, 113*t*, 116*t*, 118, 119
design of, 6, 103, 107, 187, 216n5
motivations for change in, 224–225n1
overlapping areas' advantage to developers, 152
supply/demand mismatch solution through changes in, 104, 105–106
Boundary flexibility. *See also* Annexation
advantages of, 6
BexarMet expansion, 109–110, 225n8
conditional theory's predictions, 40*t*, 111–112, 179*t*
developers' exploitation of, 152
of EBMUD, 132, 143, 144
of independent special districts, 28
interest group strategies, effect on, 171, 184–185
interlocal agreements, effect on, 48–49, 107–122, 116*t*, 117*g*, 117*t*, 120*t*, 183–184, 184, 209–211
in multiple jurisdictions, problems created by, 169–170
purposes of, 46, 224–225n1
research on, 225n9
rules on, 107–108, 110, 111, 112–113, 115, 117*g*, 120–121, 120*t*
water/land-use disputes, effect on, 140, 184, 186
Boundary review commissions, 107
Bourdeaux, Carolyn, 216n7
Buckhorn Canyon, 134, 144, 145
Building Industries Association, 145
Bureaucracies, 29, 39
Burns, Nancy, 6, 89

California
authority over land-use policy in, 131–132
designation of special districts, 27
Dougherty Valley dispute, 132–136, 140, 141*t*, 142–153, 146*t*, 170–171, 172, 182, 183
droughts in, 93, 134, 135–136, 142, 144–145, 182, 183

geographic data on water districts, 223–224n12
impact fee levels, 82, 83, 91–95, 95g, 96g, 97–100, 98t
Mello-Roos Community Facilities Act (1982), 224n15
Mitigation Fee Act (1987), 92, 224n16
moratoria on water-service connections in, 125–126
population increase in, 91, 181, 224n13, 224n19
Proposition 13, 92
Proposition 218, 92, 224n14
Rohnert Park dispute, 133m, 134–136, 163, 164t, 165–169
special district expenditures in, 5
State Water Project, 148
Tassajara Valley dispute, 132–136, 153, 154t, 155–158, 170–171, 172
water adequacy law, 128, 158, 173
water/land-use planning integration in, 129g
water metering in, 58–59
water pricing strategies in, 221n23
water shortages in, 10
water supply management, 11–12
California Building Industry Association survey, 92
California Environmental Quality Act, 131
California Statewide Database (SWDB), 204
Caps on water-service connections, 10, 81, 84–87, 125–127, 177
Case studies
  Adams Township dispute, 137–138, 137m, 158–160, 172
  California impact fees, 91–94
  Dougherty Valley dispute, 132–134, 133m, 140, 141t, 142–153, 146t, 155, 156, 157, 158, 167–168, 170–171, 172, 182, 226n6, 227n10
  East Cocalico Township dispute, 137m, 138–139, 161–162, 171, 172

Rohnert Park dispute, 133m, 134–136, 163, 164t, 165–169, 173
Tassajara Valley dispute, 132–134, 133m, 150, 151, 153, 154t, 155–158, 167–168, 170–171, 172, 173
Census of the Governments (U.S. Census Bureau), 3, 51, 52t, 53, 196, 203, 209
Citizens for Balanced Growth, 150
City government. *See* Local governments
City incorporation, 7
Climate
  cooperative relationships, effect on, 116t, 120
  election rules and, 78g
  global climate change, 10, 23, 80, 192–193
  impact fees adoption and, 93, 98t, 99, 224n21
  increasing block rates adoption and, 67, 69–72, 70t, 77–78, 77g, 78g, 79, 80
  institutional design, effect on, 53
  responsiveness, effect on, 180
  mismatch problem solution, effect on, 115
  as water supply/use predictor, 69
Climate Diagnostics Center, 204
Coate, Stephen, 43
Collective action problems, 16, 30–33, 48, 50, 218n27
Colorado River, 11, 192
Columbia River, 192
Commercial water consumption, 12g
Conditional theory of specialized governance
  assumptions of, 8–9
  on boundary contiguity's effects, 40t, 49, 171, 179t, 184
  on boundary flexibility's effects, 40t, 48–49, 112–116, 113t, 116t, 117g, 118–122, 120t, 179t, 183–184
  on election rules' effects, 40t, 44, 48, 75–79, 77g, 78g, 179t, 183
  implications of, 184–189

Conditional theory of specialized governance (cont.)
on problem severity's effects, 9, 40t, 41–46, 44t, 48, 67, 69–74, 70g, 71g, 72g, 77–79, 77g, 78g, 80, 91, 93, 97–99, 179t, 180–182
reconciliation of theories, 2, 23, 25–26, 38–39, 41
support for, 178, 179t, 180–183
test on adoption of increasing block rates, 40t, 67, 70–80, 70t, 71g, 72g, 77g, 78, 80
test on use of impact fees, 82, 90–91, 94–95, 95g, 96g
Connection fees, 149, 222n1
Conservation rates. *See* Increasing block rates
Contra Costa County Board of Supervisors, 140, 151, 152
Contra Costa County disputes (California), 173. *See also* Dougherty Valley dispute; Tassajara Valley dispute
Contracts and agreements. *See also* Cooperative relationships
benefits of, 16, 46–47
between BexarMet and SAWS, 109–110
as land-use/water dispute solution, 139–140
metropolitan reform theory on, 31
motivations for, 225n2, 225n5, 225n6
opportunities for, 114–115
public choice theory on, 35
relationship to boundary flexibility, 48–49, 107–122, 116t, 117g, 120t, 184
as supply/demand mismatch solution, 104–105, 106–107, 111–112, 121
for water supply management, 16–17, 19
Cooperative relationships. *See also* Contracts and agreements
annexation/boundary flexibility and, 48–49, 107–122, 116t, 117g, 120t, 122, 184, 225n5, 225n6, 225n9
benefits of, 18, 19, 46–48, 49, 191
between BexarMet and SAWS, 109–110
complications in systems of specialized governance, 169–173, 186–188
conditional theory on, 26, 40t, 46–49, 179t, 183–184
to cope with effects of global climate change, 193
costs of, 47–48
in growth planning, 86, 89–90, 101–102, 147, 223n10
metropolitan reform theory on, 32, 38
opportunities for, 114–115
in Pennsylvania, 136–137
political insulation and, 218n28
prior appropriation doctrine and, 13
public choice theory on, 35
as supply/demand mismatch solution, 104–105, 106–107, 111–112, 121
for sustainability, 191
Coordinated planning, 86. *See also* Contracts and agreements; Cooperative relationships
Cotati, California, 165, 167
County government. *See* Local governments

Dahl, Robert, 217n12
Dams
consequences of, 17
construction of, 10–11
environmentalists' opposition to, 135, 165–168, 175
removal of, 11, 12
Danville, California, 157
Declining block rates, 58, 59, 60, 60g, 62, 64, 221nn22–23
Demand-side policies, 2–3, 14–16, 176–177. *See also* Caps on water-service connections; Impact fees; Increasing block rates; Moratoria on

Index

water-service connections; Pricing strategies
Department of Water and Power (Los Angeles), 37, 62, 220n12
Dependent districts, 26–27, 29, 168–169, 213n1
Desalination plants, 175, 176
Developers
 benefit of water adequacy laws for, 128
 benefits from biased system, 49–50
 domination of general-purpose venues, 180, 183, 185, 186
 domination of special districts, 36–37, 38, 50, 66, 89–90, 100, 180, 185, 186
 Dougherty Valley dispute, 132–134, 133m, 140, 141t, 142–147, 146t, 152, 167–168
 goals for water supplies, 130
 identification/payment for new water sources, 127
 impact fees and, 81, 83–84, 87, 88, 90, 92–93, 99, 100, 101
 increasing block rates and, 65
 opposition to (see Antigrowth activists)
 Rohnert Park dispute, 163–169, 164t
 special district formation, 6–7, 52, 89, 186
 Tassajara Valley dispute, 132–134, 133m, 154t, 155–158, 167–168
Development. *See* Growth and development
Development fees. *See* Connection fees; Impact fees; Land-use exactions
Dillon's Rule, 213n2
Doig, Jameson, 218n28
Domestic water consumption, 12g, 14
Dougherty Valley dispute
 EBMUD and, 140, 141t, 142–147, 146t, 149, 150, 151–152, 183
 impact on Tassajara Valley project, 155, 156, 157
 issue salience/problem severity in, 144, 145–146, 149, 151–152, 172, 182, 183
 legislation requiring consideration of water availability, 226n6
 location, 133m
 specialized governance's effect on, 167–168, 170–171
 Zone 7 and, 141t, 147–153, 183
Drinking water policy. *See* Connection fees; Impact fees; New local politics of water; Pricing strategies; Sustainability policies
Droughts
 adoption of increasing block rates during, 62
 in California, 93, 134, 135–136, 142, 144–145, 148, 182, 183
 global climate change and, 192
 impact fees use, effect on, 94, 224n22
 moratoria on water-service connections during, 126
 preparation for and water adequacy laws, 128
 in San Antonio area, 123
 in Southwest, 73
 water demand variations due to, 225n3
 water shortages during, 9, 10
DSRSD. *See* Dublin San Ramon Services District
Dublin, California, 147, 149
Dublin San Ramon Services District (DSRSD), 140, 147, 149–150, 151, 152, 153

East Bay disputes. *See* Dougherty Valley dispute; Tassajara Valley dispute
East Bay Municipal Water District (EBMUD)
 concerns about water supply adequacy, 134
 Dougherty Valley dispute, 140, 141t, 142–147, 146t, 149, 150, 151–152, 183

East Bay Municipal Water District (cont.)
  effort to consider water availability in land-use decisions, 226n6
  policy on expansion of boundaries, 132
  Tassajara Valley dispute, 153, 155–157
East Cocalico Planning Commission, 161–162
East Cocalico Township dispute, 137*m*, 138–139, 161–162, 171, 172
East Cocalico Water Authority, 161–162
Economic efficiency
  empirical literature on, 36, 37
  impact fees and, 87
  public choice theory on, 33–34, 35
  through increasing block rates, 23, 61, 64, 78–79
Edwards Aquifer Authority, 123, 175
Eel River, 227–228n20
EIRs. *See* Environmental impact reports
Elections
  approving UGB, 227n16
  for board of EBMUD, 143–144, 145–146, 146*t*, 153, 227n8
  for board of SCWA, 169
  for board of Zone 7, 148
  conditional theory on, 40*t*, 42–44
  metropolitan reform theory on, 38
  public choice theory on, 34–35, 38
  of special district officials, 31, 32, 75, 76, 77*g*, 179*t*
  ward vs. at-large, 75, 76, 78, 78*g*, 222n31
Electricity pricing plans, 59
EMBUD. *See* East Bay Municipal Water District
Endangered species, 16, 123
Environmental impact reports (EIRs)
  California requirement for development, 131–132
  for Dougherty Valley project, 140, 142, 143, 149
  for Tassajara Valley project, 153, 155–156, 157
Environmentalists
  Dougherty Valley dispute and, 133, 143–146, 148, 150, 152
  impact fees, support for, 84, 100
  increasing block rates, support for, 55
  in Rohnert Park, 165
  in Sonoma County, 134, 135–136
  Tassajara Valley dispute and, 133, 156–157
  voter support for development, influence on, 228n21
  water supplies goals, 17, 130, 177
  water use restrictions and, 16
Environmental Protection Agency. *See* U.S. Environmental Protection Agency
Environmental regulation, 11, 59, 60, 80, 175
Environmental review process, 131–132, 142, 165, 188
EPA. *See* U.S. Environmental Protection Agency
Erie, Steven, 217n17
Ernst & Young surveys, 219n6
Evapotranspiration rates, 192

Federal government
  assistance for large-scale infrastructure projects, 3, 10–11
  cutbacks on funding for local public services, 83
  regulation of endangered species, 16
  special district formation, effects on, 6
  sustainability policies and, 191–192
  water quality/efficiency standards, 14–15, 16, 18
Fire protection, 5, 44, 56, 57, 218n24
Flat-fee rate system, 58, 60*g*
Florida, 10, 11, 12, 73, 86, 175, 224n20
Fort Collins, Colorado, 62–63
Foster, Kathryn, 7, 36, 107, 185

Fragmentation of authority. *See also* Special districts
  accountability/visibility of governing units and, 31
  as collective action problem, 218n27
  cooperative relationships in, 86, 106, 169–173, 178
  debate over, 30, 100
  impact fees use, effect on, 82
  land-use issues and, 88–89, 124–125, 131–136, 133*m*, 140, 141*t*, 142–153, 146*t*, 154*t*, 155–162, 226n6, 227n10
  local water review processes, effect on, 128
  metropolitan reform theory on, 31, 38
  negative externalities due to, 32, 178
  public choice theory on, 33–35
  sustainability initiatives and, 191
  water/land-use planning integration and, 128–130, 129*g*
Fresno, California, 58
Functional specialization. *See* Special districts; Water districts

GAO. *See* U.S. Government Accountability Office
General-purpose venues. *See also* Federal government; Local governments; State governments
  ability to address complex problems, 45
  access to officials, 35, 66, 79
  adoption of increasing block rates, 62–63, 66, 69–74, 70*t*, 71*g*, 72*g*, 78–79, 78*g*, 80
  annexation by, 105–106
  coordination with special districts, 88, 89, 101–102
  developer domination of, 49–50, 180, 183, 185, 186
  impact fees use, 82, 90, 94–95, 95*g*, 96*g*, 97–100, 98*t*, 100–101
  responsiveness of, 39, 42–45, 44*g*, 45, 183
  ward vs. at-large elections, effect of, 76–77, 78, 78*g*
  water/land-use planning integration and, 124, 128–130, 129*g*, 159, 163–169, 164*t*
Georgia, 11, 18, 19, 73
Gerber, Elisabeth, 228n21
Gleick, Peter H., 17
Global climate change, 10, 23, 80, 192–193
Global Water Partnership, 17
Gómez-Ibáñez, José, 82–83, 84
Government Accountability Office. *See* U.S. Government Accountability Office
Great Lakes, 11, 192
Greenhouse-gas emissions, 192
Groundwater aquifers
  aquifer storage and recovery projects, 176
  depletion of, 72–73, 80, 175
  development of cooperative institutions for management of, 16–17
  overdrafting of, 10, 14, 36, 104–105, 215n28
  regulation of extraction from, 13–14
  as source of public water, 12*g*
Growth and development
  Adams Township dispute, 137–138, 137*m*, 158–160
  in California, 91, 131–132, 133*m*, 139–140
  cost distribution of, 87–88
  Dougherty Valley dispute, 132–134, 133*m*, 140, 141*t*, 142–153, 146*t*, 226n6, 227n10
  East Cocalico Township dispute, 137*m*, 138–139, 161–162
  environmentalists' opposition to, 133, 135–136, 143–146, 148, 150, 152, 156–157, 167–168, 177, 228n21
  impact fees and, 57, 81, 82, 84–87, 93, 94–95, 95*g*, 97–99, 181
  measurement of, 92–93

Growth and development (cont.)
  metropolitan reform theory on, 36–37
  moratoria/restrictions on water-service connections and, 81, 125–126, 127, 177
    opposition to, 82, 83–84, 90, 185–186
    in Pennsylvania, 136–137, 137*m*, 139–140
  relationship to water policy, 18, 23, 44–45, 105, 123–124, 177, 182 (*see also* Impact fees)
  Rohnert Park dispute, 133*m*, 134–136, 163, 164*t*, 165–169
  special districts' effect on, 36–37, 180, 185, 217n12
  special districts' role in planning/management of, 1, 88–89
  supply/demand mismatch from, 103, 105
  sustainability policies and, 190, 191
  Tassajara Valley dispute, 132–134, 133*m*, 153, 154*t*, 155–158
  water systems, effect on, 82, 85
Growth machine, 49–50, 90, 180, 183, 185, 186. *See also* Developers
Growth policy. *See* Connection fees; Environmental impact reports (EIRs); Impact fees; Land-use exactions; Land-use management; Moratoria on water-service connections; Urban-growth boundaries (UGBs); Water adequacy laws

Hanak, Ellen, 93
Heikkila, Tanya, 218n21
Hooghe, Liesbet, 107
Houston, Texas, 10, 36, 37, 72–73

Impact fees
  benefits of, 100
  in California, 91–92
  climate's effect on, 99
  definition of, 222n1
  disadvantages of, 87
  empirical literature on, 86–87, 223n7
  as form of growth policy, 81, 84–87, 100–101, 181, 223n4
  income's effect on, 99
  influence of governing structure on use of, 91–95, 95*g*, 96*g*, 97–102, 98*t*
  intergovernmental coordination and, 86, 88–89, 101–102
  legislation regulating use of, 224n16
  motivations for, 23, 82–84, 85, 86, 219n3
  opposition to, 84, 87, 88, 90, 99, 100
  relationship to growth and development, 93, 94–95, 95*g*, 96*g*, 97–99
  reliance on, 87–88
Increasing block rates
  adoption of, 60, 64, 177, 221nn22–23
  climate's effect on adoption of, 69–71, 70*t*, 77*g*, 78–79, 78*g*
  determining rates by income, 65, 220n11
  election rules' effect on adoption of, 74–79, 77*g*, 78*g*, 80, 222n31
  as inducement to conservation, 23, 55, 57, 61, 64, 80, 220n8, 221n15
  opposition to, 57–58, 59, 62–63, 64, 65–66, 180, 220n12, 220n13
  public benefits of, 23, 55, 58, 61–62, 64–65, 78–79, 219–220n7
  revenue variability, 61
  specialized governance's effect on adoption of, 66–74, 70*t*, 71*g*, 72*g*, 195–198, 199*t*, 200–201
Indianapolis, Indiana, 18
Institutions, 39, 41
Intergovernmental coordination. *See* Cooperative relationships
Intergovernmental Panel on Climate Change, 192
Interlocal cooperation. *See* Contracts and agreements; Cooperative relationships

Inverted rates. *See* Increasing block rates
Ipswich River, 10
Irrigation, 9, 12*g*, 14, 15, 56, 176
Isett, Kimberly, 218n21
Issue bundling, 43–44, 76
Issue salience
  in Dougherty Valley dispute, 144, 145–146, 149, 151–152, 172, 182, 183
  general-purpose governments' response to, 183
  increasing block rates adoption, effect on, 67, 73–74
  land-use disputes and, 134, 138, 172
  in Pennsylvania disputes, 172, 182
  problem severity and, 42, 69, 181–183
  public good provisions, effect on, 71–74
  public opinion measures of, 218nn22–23
  responsiveness, effect on, 42–45, 180, 183
  special districts' response to, 183
  water/land-use disputes, effect on, 140

Jones, Victor, 32, 89

Lancaster County, Pennsylvania, 137*m*, 138–139, 161–162
Land-use disputes. *See* Growth and development
Land-use exactions, 81, 83, 84–87, 92, 223n9. *See also* Impact fees
Land-use management. *See also* Growth and development; Impact fees; Land-use exactions; specific dispute over development
  impact fees and, 81–82, 84–87, 100–101, 180–181
  influence on growth and development, 23
  integration with water planning, 125–130, 176, 180, 226n6
  opportunities for public involvement in, 131, 182, 183, 184
  policy spillovers and, 32
  relationship to water supply, 16, 19, 44–45, 127, 175
  restrictions on building or new connections, 10, 81, 85, 125–126, 177
  voter impact on, 228n21
Lifeline plans, 62, 220n14
Lin, Ben, 216n7
Livermore, California, 148, 150–151
Local governments. *See also* Cooperative relationships; General-purpose venues; Growth and development; Land-use management
  annexation rules, 113, 113*t*
  authority of, 213n2
  authority over land-use management, 88, 90
  constraints on, 41–42
  functions/boundaries of, 28, 29
  numbers of, 4
  utility service area expansion, 105–106
Logrolling, 25, 43
Los Angeles, California, 37, 62, 220n12

Marks, Gary, 107
Mead, Lake, 10
Metering, 57, 58–59, 173, 176
Metropolitan reform theory, 30–33, 35–37, 38–39, 48, 89, 183
Metropolitan Water District (California), 37, 217n17
Milwaukee, Wisconsin, 18
Moratoria on water-service connections, 10, 81, 85, 125–126, 177

National Climatic Data Center (NCDC), 196, 204
National Research Council Water Science and Technology Board, 215n23

NCDC. *See* National Climatic Data Center
New local politics of water, 2–3, 9, 18–19, 51, 67, 85, 127, 175, 177, 178, 189–193
New regionalism, 33
New York, 4, 5, 193
North Carolina, 10, 11

OECD. *See* Organization for Economic Cooperation and Development
Ogallala Aquifer, 192
Orange County Water District (California), 175
Organization for Economic Cooperation and Development (OECD), 219n4
Orme, Tennessee, 10
Ostrom, Elinor, 107

Pacific Institute, 17
Palmer Hydrological Drought Index (PHDI), 205
Peiser, Richard, 216n7
Penngrove, California, 165–166, 167, 168
Pennsylvania
 Adams Township dispute, 137–138, 137*m*, 158–160, 182
 East Cocalico Township dispute, 137*m*, 138–139, 161–162
 Municipalities Planning Code, 136
 special district expenditures in, 5
 special district governance in, 27, 172
 state oversight of planning activities, 136–137
 water issue salience in, 172, 182
 water supplies distribution in, 173
Peterson, Paul, 50
PHDI. *See* Palmer Hydrological Drought Index
Phillips, Justin, 228n21
Pittsburgh, Pennsylvania, 137
Pleasanton, California, 148, 227n10

Polycentrists, 30, 33–35
Population growth/redistribution. *See also* Growth and development
 in California, 91, 132, 135
 cooperative relationships, effect on, 116*t*
 demand/capacity mismatch and, 10, 103, 115, 175
 depopulation, 103–104
 impact fees, effect on, 94–95, 95*g*, 96*g*, 97
 in Pennsylvania, 136, 137
 impact fee use, promotion of, 86–87
 rate structure and, 228n3 (app. 1)
 in San Antonio, 123
Porter, Douglas, 216n7
Potomac River, 11
Power plants, 12*g*, 13
Prairie Waters Project, 176
Precipitation. *See* Climate
Pricing strategies
 for agricultural water supplies, 219n2
 as conservation approach, 57, 221n15
 declining block rate, 58, 59, 60, 60*g*, 62, 64, 221nn22–23
 flat-fee system, 58, 60*g*
 impact fees, 23, 81–84
 increasing block rate, 23, 58, 60–65, 60*g*, 80, 180, 220nn7–8, 220nn11–13
 lifeline rates, 62, 220n14
 as policy instrument, 55, 57–58, 219nn2–3
 price increases, 219n4
 redistribution of income through, 23, 55, 57, 61, 62, 65
 uniform rate, 58, 59, 60, 60*g*
Prior appropriation doctrine, 13
Private goods, 56–58
Private water companies, 18–19, 57, 86, 215n25
Problem context, 25–26, 46–49, 79, 115, 140. *See also* Issue salience; Problem severity

Problem severity
  in California, 91
  conditioning effect of, 41–46, 180–183
  in Dougherty Valley dispute, 144–146, 149, 151–152, 172, 182
  cooperative relationships, effect on development of, 48, 116t, 171
  global climate change and, 193
  impact fees, effect on, 92, 96g, 97–99, 100, 102
  increasing block rates adoption, effect on, 67, 70t, 71–74, 71g, 72g, 77, 80
  interlocal agreements participation, effect on, 119–120
  issue salience and, 181–183
  mismatch problem solution, effect on, 115, 116t
  responsiveness, effect on, 9, 40t, 42, 44g, 45–46, 172, 179t, 180, 182–183
  water/land-use disputes, effect on, 140
Progressive rate structures. *See* Increasing block rates
Public authorities. *See* Special districts
Public choice theory, 7–8, 23, 33–35, 36, 38–39, 48, 66–67, 90, 183, 217n18
Public goods, 2, 19, 23, 26, 32, 33, 34, 35, 46, 56–58, 68, 71, 73–74, 79, 80, 87, 180, 183, 185, 191
Public hearings, 108, 131, 165–166, 226n17
Public Utilities Regulatory Policies Act (1978), 59
Public water supply, 12–14, 12g, 14g, 15–16, 15g, 17, 20, 214n13, 214n17

Raftelis surveys, 219n6
Rationing, 61, 176, 221n15
Responsiveness
  conditional theory on, 20, 23, 40t, 41–46, 178, 179t, 180, 181, 183

definition of, 49, 217n11
election rules' effect on, 39, 40t, 44, 74–79, 77g, 78g, 80, 179t, 183
empirical literature on, 36, 37
intergovernmental coordination and, 101–102
metropolitan reform theory on, 31–32, 34, 38, 39
new local politics of water and, 3, 18, 19, 49, 51, 178
of private water companies, 18, 19
problem severity/issue salience's effect on, 40t, 42–43, 45–46, 69–74, 70t, 71g, 72g, 101, 179t, 180–182, 222n30
public choice theory on, 2, 8, 34, 38, 39, 66
specialized governance's effect on, 39, 41–45, 44g, 66–67, 69–74, 70t, 71g, 72g, 80, 88–91, 94–95, 95g, 96g, 97–102, 98t, 178, 180, 181, 186
value of, 188–189
Restrictions on water use, 10, 16, 17, 93, 189
Retail water deliveries, 73, 214n16
Revenue bonds, 6–7, 9, 213n6
Richmond v. Shasta Community Services District (California), 224n14
Riparian doctrine, 13
Roanoke River, 11
Rohnert Park dispute, 133m, 134–136, 163, 164t, 165–169, 173
Russian River dam, 135, 165, 168

Sacramento, California, 58–59
Safe Drinking Water Information System (EPA), 204
Salience of issues. *See* Issue salience
Saltwater intrusion, 73
Salyer Land Company v. Tulane Lake Water Storage District, 31, 217–218n20
San Antonio, Texas, 109–110, 123

San Antonio Water System (SAWS), 109–110, 111, 123–124, 175, 226n1
SAWS. *See* San Antonio Water System
SCWA. *See* Sonoma County Water Agency
Sierra Club, 135, 143, 150
Smith, Robert, 216n7
Sonoma County dispute, 133*m*, 134–136, 163, 164*t*, 165–169, 173
Sonoma County Water Agency (SCWA), 134, 135, 165, 168, 227–228n20
South Platte River, 176
Special districts. *See also* Water districts; specific water district by name
  accountability of, 31, 33, 34, 43, 75–76, 183–184, 185
  appointment of district officials, 31, 32, 38, 39, 48, 75–76, 80, 115, 183
  authority of, 5, 26–28
  boundaries of (*see* Boundaries; Boundary flexibility)
  cooperative relationships (*see* Contracts and agreements; Cooperative relationships)
  defining characteristics, 3, 23–30, 216nn2–3
  dependent vs. independent, 26–27, 29, 168–169, 213n1
  efficiency of, 34, 35, 36, 37, 217n12, 217n15, 217n17
  election of district officials, 31, 32, 34–35, 38, 39, 75, 76, 78, 78*g*, 80, 115, 119, 143–144, 145–146, 146*t*, 148, 153, 169, 227n8
  elimination of, 33, 110–111
  expenditures by, 4, 5–6, 20, 21*m*, 36, 37
  formation of, 6–7, 9, 51–52, 89, 107, 186
  functions of, 1, 2, 3, 4–5, 36–37
  issue expertise, 1, 25, 29, 41, 45, 75, 80
  limitations of, 186
  locations of, 5, 22*m*
  proliferation of, 1, 2, 4, 4*g*, 6
  responsiveness, 31–32, 34, 38, 41–45, 44*g*, 160, 178, 179*t* (*see also* Responsiveness)
  visibility of, 7, 8, 25, 29–30, 31–32, 36, 43–44, 183, 186, 216n7, 218n21, 218n28
Spillovers, 32, 35, 47, 103, 106, 167, 169–170, 171
State governments
  assistance for large-scale infrastructure projects, 3, 10–11
  cutbacks on funding for local public services, 83
  regulation of land-use policies, 19, 85
  special district formation and design, effect on, 5, 6, 29, 53, 107–108, 110, 111, 112, 113*t*, 187
  sustainability policies and, 191–192
  water adequacy laws, 127–129
Stein, Robert, 112, 225n5
Stockton, California, 18
Supply-side solutions, 2–3, 10–11, 16, 82, 175–176. *See also* Dams; Desalination plants; Water sources
Supreme Court, 31, 83, 217–218n20
Surface water, 12*g*, 13, 80, 98*t*, 100
Sustainability policies, 18, 189–193
SWDB. *See* California Statewide Database

Tampa, Florida, 10, 73, 175
Tassajara Valley dispute
  Dougherty Valley dispute and, 150, 151
  issue salience/problem severity in, 172
  location, 133*m*
  specialized governance's effect on, 170–171
Tassajara Valley Property Owners' Association (TVPOA), 153, 154*t*, 155
Temperatures. *See* Climate
Teodoro, Manuel, 222n30

Texas
  BexarMet expansion, 109–110, 111, 123–124, 182
  desalination plants, use of, 175
  Edwards Aquifer Authority establishment, 123
  overtapping of groundwater aquifers, 10, 36, 72–73
  progrowth outlook in, 36–37
  special district expenditures in, 5
  water districts, reliance on, 53
Thermoelectric power production, 12g, 13, 15
Tiebout, Charles, 33, 35–37
Toilet to tap projects, 176
Tufte, Edward, 217n12
TVPOA. *See* Tassajara Valley Property Owners' Association

UGBs. *See* Urban-growth boundaries
Uniform rates, 58, 59, 60, 60g, 65
Urban-growth boundaries (UGBs), 135, 138, 140, 163, 227n16
U.S. Army Corps of Engineers, 10–11
U.S. Bureau of Reclamation, 10–11
U.S. Census Bureau, 3, 27, 51, 52t, 53, 195, 196, 203, 209, 216n2
U.S. Environmental Protection Agency, 18, 85, 204, 219n5
U.S. Government Accountability Office (GAO), 83

Virginia, 11
Voting Rights Act, 6

Wagner, Richard, 217n15
Ward elections, 75, 76, 78, 78g, 222n31
Washington (state), 213n6, 214n15
Washington, D.C., 11
Water
  affordability of in U.S., 17, 18, 219nn5–7
  as a good, 56–58
  as a natural monopoly, 20
  residential consumption, 220n11
  social/cultural meaning of, 188–189
  true marginal cost of, 219–220n7
  uses of, 12–13, 12g
Water adequacy laws, 127–129, 158, 173
Water availability
  development and, 172–173, 182
  factors affecting, 181–182
  global climate change and, 192–193
Water demand
  in cities, 214n17
  droughts and, 9
  factors increasing, 10, 56–57, 103, 175
  income and, 62, 220n11
  price structure response, 61
  reduction strategies (*see* Demand-side policies; Impact fees; Sustainability policies)
  seasonal variations, 225n3
  supply mismatch with, 103–105
  trends, 14g
Water distribution systems. *See also* New local politics of water; Special districts; Water demand; Water districts
  challenges to, 79–80, 175, 189–193
  cost structures of, 56
  debt, 114, 116t, 119
  development of, 56–57, 79, 173
  financing of expansion, 82
  history of, 2–3, 10–12, 17, 56–59, 175–176
  integration with land-use planning, 125–130, 176, 180, 226n6
  responsibilities of, 56
  shift from private to public governance of, 80
Water districts. *See also* specific water district by name
  boundary flexibility (*see* Boundary flexibility)
  definition of, 215n26
  establishment of, 9, 51–52
  expenditures by, 20, 21m

Water districts (cont.)
  numbers and locations of, 20, 22*m*, 52*t*, 53
  rules for boundary changes, 110, 111, 112
  staff size, 46
Water pollution, 9, 15*g*, 16, 103, 104–105, 190
Water-pricing strategies. *See* Pricing strategies
Water rate structures. *See* Pricing strategies
Water-resource managers, 49–51, 55, 63–64, 81, 138–139, 177
Water rights, 11, 13–14
Water-sewer districts, 29
Watershed management, 16–17, 177
Water sources. *See also* Dams; Groundwater aquifers; Surface water
  aquifer storage and recovery projects, 176
  competition for access to, 2–3, 17
  degradation of, 103, 104–105
  desalination, 175, 176
  global climate change and, 192–193
  large storage projects, 2–3, 10–12, 17, 175–176 (*see also* Dams)
  limitations of, 9–10, 13–14, 175
  for public water systems, 12*g*, 13, 173
  water rights and, 11, 13–14
Water transport issues, 10, 11–12, 124, 127, 149–150, 152, 153, 190–191
Water treatment, 214n17
Water-use restrictions, 10, 16, 17, 93, 189
Weber, Max, 29
Weber, Warren, 217n15
West View Water Authority, 138, 158, 159–160
Wheeling, 127
Wilson, James Q., 221n17
Wilson, Woodrow, 29

Wlezien, Christopher, 218n22
World Water Forum (2009), 215n22

Zone 7 Water Agency, 141*t*, 147–153, 183

American and Comparative Environmental Policy
Sheldon Kamieniecki and Michael E. Kraft, series editors

Russell J. Dalton, Paula Garb, Nicholas P. Lovrich, John C. Pierce, and John M. Whiteley, *Critical Masses: Citizens, Nuclear Weapons Production, and Environmental Destruction in the United States and Russia*

Daniel A. Mazmanian and Michael E. Kraft, eds., *Toward Sustainable Communities: Transition and Transformations in Environmental Policy*

Elizabeth R. DeSombre, *Domestic Sources of International Environmental Policy: Industry, Environmentalists, and U.S. Power*

Kate O'Neill, *Waste Trading among Rich Nations: Building a New Theory of Environmental Regulation*

Joachim Blatter and Helen Ingram, eds., *Reflections on Water: New Approaches to Transboundary Conflicts and Cooperation*

Paul F. Steinberg, *Environmental Leadership in Developing Countries: Transnational Relations and Biodiversity Policy in Costa Rica and Bolivia*

Uday Desai, ed., *Environmental Politics and Policy in Industrialized Countries*

Kent Portney, *Taking Sustainable Cities Seriously: Economic Development, the Environment, and Quality of Life in American Cities*

Edward P. Weber, *Bringing Society Back In: Grassroots Ecosystem Management, Accountability, and Sustainable Communities*

Norman J. Vig and Michael G. Faure, eds., *Green Giants? Environmental Policies of the United States and the European Union*

Robert F. Durant, Daniel J. Fiorino, and Rosemary O'Leary, eds., *Environmental Governance Reconsidered: Challenges, Choices, and Opportunities*

Paul A. Sabatier, Will Focht, Mark Lubell, Zev Trachtenberg, Arnold Vedlitz, and Marty Matlock, eds., *Swimming Upstream: Collaborative Approaches to Watershed Management*

Sally K. Fairfax, Lauren Gwin, Mary Ann King, Leigh S. Raymond, and Laura Watt, *Buying Nature: The Limits of Land Acquisition as a Conservation Strategy, 1780–2004*

Steven Cohen, Sheldon Kamieniecki, and Matthew A. Cahn, *Strategic Planning in Environmental Regulation: A Policy Approach That Works*

Michael E. Kraft and Sheldon Kamieniecki, eds., *Business and Environmental Policy: Corporate Interests in the American Political System*

Joseph F. C. DiMento and Pamela Doughman, eds., *Climate Change: What It Means for Us, Our Children, and Our Grandchildren*

Christopher McGrory Klyza and David J. Sousa, *American Environmental Policy, 1990–2006: Beyond Gridlock*

John M. Whiteley, Helen Ingram, and Richard Perry, eds., *Water, Place, and Equity*

Judith A. Layzer, *Natural Experiments: Ecosystem-Based Management and the Environment*

Daniel A. Mazmanian and Michael E. Kraft, eds., *Toward Sustainable Communities: Transition and Transformations in Environmental Policy*, 2nd edition

Henrik Selin and Stacy D. VanDeveer, eds., *Changing Climates in North American Politics: Institutions, Policy Making, and Multilevel Governance*

Megan Mullin, *Governing the Tap: Special District Governance and the New Local Politics of Water*